Iosef Lvovich Ostrovsky, my father

Boris Ostrovsky

CONVULSIONS OF THE PLANET EARTH

The Bermuda Triangle Changes Its Address

AUSTIN MACAULEY PUBLISHERS™

LONDON * CAMBRIDGE * NEW YORK * SHARJAH

Ordering Information
Quantity sales: Special discounts are available on quantity purchases by corporations, associations, and others. For details, contact the publisher at the address below.

Publisher's Cataloging-in-Publication data
Ostrovsky, Boris
Convulsions of the Planet Earth

ISBN 9798889103646 (Paperback)
ISBN 9798889103653 (ePub e-book)

Library of Congress Control Number: 2023915695

www.austinmacauley.com/us

First Published 2023
Austin Macauley Publishers LLC
40 Wall Street, 33rd Floor, Suite 3302
New York, NY 10005
USA

mail-usa@austinmacauley.com
+1 (646) 5125767

The author expresses his deep gratitude to

- Lev Eppelbaum, Professor of the Department of Geophysics at Tel Aviv University, for consultations and valuable advice in the process of writing this book;

- Valeria Mednikov, philologist and mathematician, for great help in translating the book and communicating with the Publishing House.

Table of Contents

BORIS OSTROVSKY AND HIS BOOK

Having opened this book, some readers will frown – again he is being fed reading about the Bermuda Triangle. The annoyance of the reader can be understood – after all, the mysteries of the notorious Triangle are on a par with publications about the UFO phenomenon and the intrigues of insidious aliens. Interest in such stories faded over time because no evidence of their reality was presented. Except for one indisputable fact: planes and ships really disappeared without a trace in the Bermuda Triangle. Why is the problem of natural science, which has existed for more than half a century, to this day remains "the greatest mystery of our time"? Yes, because this problem has been exploited for years by dilettantes, whose main goal was to capture the imagination of the reader and gain popularity. Their speculative versions, devoid of any scientific basis, wandered from one publication to another for so long that, in the end, they formed a skeptical attitude toward the topic in the readership.

The book by Boris Ostrovsky, which you, dear reader, are now holding in your hands, is fundamentally different from the works of his predecessors. Having many years of experience in scientific work behind him, the author (Doctor of Medical Sciences) undertook to study and systematize the extensive factual material presented in the documents of the Bermuda disaster investigation commissions, in the reports of eyewitnesses of the events – pilots and sailors.

At the same time, the recurrence of such mysterious phenomena as light effects in the sky and under water, a violation of the functions of navigation instruments, the appearance of UFOs in the area of the incident, the disappearance of ships and aircraft, etc., could not help but attract attention. Ostrovsky was the first to succeed in explaining the nature of these "Bermuda Phenomena" from the standpoint of the "classical" sciences – physics, geology, oceanography, etc. The results of many years of research led the author to the conclusion that all "Bermuda phenomena" are due to geophysical causes, mainly active geodynamic processes at the bottom of the Northwest Atlantic.

In the 21st century, Bermuda phenomena began to be recorded in other seismically active regions of the globe, which the author demonstrated in the analysis of man-made disasters that occurred in East and Southeast Asia. Thus, he brought the topic far beyond the boundaries of the Bermuda Triangle itself, giving it a planetary scale.

Ostrovsky's explanation of the nature of the "Bermuda phenomena" by specific geophysical processes does not contradict the generally accepted principles of geology and can be regarded as a scientific hypothesis. The book has been repeatedly republished by the leading Russian publishing house AST. Written in the popular science genre, it is of interest not only to specialists but also to a wide range of readers. As "RU-Books" (an electronic library of scientific literature) notes, "Boris Ostrovsky's book is read in one breath like a famously twisted detective story."

Lev Eppelbaum, Professor
Department of Geophysics,
Tel Aviv University

Part I
Lights Over the Ocean

*"When I was a little boy I lived in an old house, and
legend told us that a treasure was buried there…
it cast an enchantment over that house. My home
was hiding a secret in the depths of its heart…"*

Antoine de Saint-Exupery, The Little Prince

1.1. A Soul-Chilling Chronicle

The day contained no omen of anything sinister. Or, moreover, anything so frightening its echoes would reverberate to all corners of the Earth. Still, even a bare-bones chronicle of that day chills the soul. Here it is…

December 5, 1945. 14 hours 10 minutes

Five bomber planes, comprising the training mission designated as Flight 19, take off from the US Navy airbase at Fort Lauderdale, situated on the eastern coast of Florida. They follow a northeast course to the Sargasso Sea. There, near the small island of Bimini, they are to conduct maneuvers. The time allotted for this training flight is two hours. Commanding Flight 19 is Lieutenant Taylor, an experienced pilot instructor. Neither are the rest of the crew novice aviators. The weather, according to the pilots, is "ideal." Having dropped their bombs on the training target, the planes set their course for home.

15 hours 15 minutes

The control center at Fort Lauderdale receives troubling radio transmissions.

Lt. Taylor: "Calling control. We have a dangerous situation. It looks like we're off course. We cannot see land…Repeat. We cannot see land…"

Control station: "Your location?"

Lt. Taylor: "We can't determine where we are. It looks like we've gone astray."

Control station: "Hold your course west."

Lt. Taylor: "We cannot determine which way is west. Something is wrong…We can't establish our direction. And the sea looks strange, somehow…"

The dispatchers onshore are startled by the unusual communications from Lt. Taylor. Sudden radio static disrupts contact between the base and the pilots. But along the coast there continue to be interceptions of scraps of radio exchange between the pilots, among them the following:

On all five planes, navigational instruments have failed. ("The compass needles are spinning like crazy." "The compasses show different directions.")

Fuel remaining on the planes is only enough for 100 km, with no land in sight.

The alarm spreads up and down the coast. Control center dispatches planes to meet up with distressed Flight 19. One of the rescue planes flies out from the military base at Banana River. This is a twin-engine Martin PBM Mariner flying boat, bearing 13 para-troopers. The specially trained group is headed by Officer Cone.

The last communiqués to be heard from lost Flight 19 are abrupt, desperate exclamations: "We're totally disoriented…We're entering white water…"

Shortly afterward, the Martin Mariner search plane makes itself heard: at a height of 1,800 meters the plane has found itself in a zone of strong turbulence. This is the last message received from the Martin Mariner.

During the first hour of the search gigantic waves are observed from the air. The distances between waves are so great that, as later recounted by the pilots of the search planes, entire airports could be situated within them.

The air search is interrupted with the onset of darkness. But military ships plow the waters throughout the night, illuminating them with their powerful searchlights in the hope of taking aboard any pilots from Flight 19 who may have ended up in the ocean. Or at least finding some traces of the catastrophe. But in vain…

December 6. 8 hours 00 minutes

The search continues. It now involves 240 aircraft (with an additional 67 supplemental planes on aircraft carriers such as the USS *Solomon*s ready to take off at any moment), several submarines, and hundreds of private planes, yachts, and boats. Later on, aid is supplied by planes from the airport at Banana River, in addition to ships from the British fleet. This morning the ocean is calm, which allows pilots to fly low over the surface. And on it goes till the very evening.

Successive observations over an area of 100 thousand square kilometers of land and water comprising the western Atlantic, Caribbean Sea, Gulf of Mexico, Florida and adjacent islands, yield nothing.

1.2. Riddles of the Bermuda Triangle

Let's say that a person is walking along the street and suddenly falls. Certainly, there's nothing unusual about this, anyone can slip. But what if on that same even surface, on that very same day, seven people slip and fall? Hmmm…well, there's still nothing supernatural about this. Such coincidences, though rare, do happen. However, if on that same even surface seven people slip and fall simultaneously, a riddle now arises, demanding explanation…

Isn't the Bermuda Triangle an analog to this situation?

"Both ships and planes experience disasters here."

"This happens in other regions of the planet as well!"

"Yes, but in the Bermuda Triangle these disasters take place in unexplained circumstances."

"It's not always possible to establish the causes of disasters on other air and marine routes around the world."

"But in the Bermuda Triangle these disasters occur more often."

"Well, it's possible that the natural peculiarities of the region create difficulties for pilots and navigators."

"Only in part: in the Bermuda Triangle ships and planes sometimes vanish even in good weather."

"Bermuda Triangle"—where did this title come from, thrilling people's imaginations and becoming "the sensation of the age"? Over time, this region of disappearances began to emerge as a clearer picture, situated between the islands of Bermuda, the southern tip of Florida, and the Island of Puerto Rico. The outlines of this region acquired the shape of a triangle.[1]

---

[1] Although the major portion of the Triangle occupies latitude close to the equator, it is conventionally spoken of as a region lying in the North Atlantic. The Atlantic Ocean lies on both sides of the Equator that divides our planet into Northern and Southern hemispheres; the Bermuda Triangle is situated in the Northern Hemisphere.

In 1950, Associated Press correspondent Edward Van Winkle Jones made a survey of air and marine disasters in this region. But the title "Bermuda Triangle" was first given it by Vincent Gaddis, who published a sensational essay in the journal *Argosy* in 1964.

Even with a fleeting acquaintance with the history of disasters in the Northwest Atlantic, it's possible to note that they are extraordinary, that in addition to established factors causing these tragedies at sea, on this patch of the planet some other kind of powers are at play.

Let's say that a dangerous situation has arisen on a vessel or plane. In these cases, an SOS signal is broadcast by radio from on board, along with the call signals of the vessel or plane, and its coordinates. The signals are received onshore, on other marine vessels and aircraft, and rescue crews stream toward the point of distress. These timely search responses allow many lives to be saved.

But in the Bermuda Triangle all of this happens somehow differently: the SOS signals often don't get through from here. The danger is detected only suddenly, and the catastrophe unfolds so rapidly that people don't have time to undertake any measures. The impression is that vessels and planes vanish in the Bermuda Triangle abruptly, as though they've plummeted into nothingness...

It's generally considered that accounts of mysterious shipwrecks in the Northwest Atlantic date from 1830. But we may surmise that vessels began to disappear much earlier in this part of the planet; in 1550, on his geographical map of the world, Giacomo Gastaldi labeled the islands of Bermuda "Devil's Islands."

Over the course of three centuries, the Spanish plied their ships full of treasure along the coast of Florida. This was a particularly convenient route, since the Gulf Stream with its 6-knot speed enabled faster voyages by sea. Many of these ships never again saw land. The waters off southern Florida became a cursed place for Spanish sailors; in the end, the region was christened "Hell's Circle."

With the development of shipping came the rapid growth of stock-trade and insurance companies. The systematic loss from ships failing to return home began to mount, and ship owners increasingly began to pose the question: why did misfortunes often occur precisely in this area of the Atlantic? In those long-ago times, there did not yet exist a multi-faceted weather service, so the losses could be attributed to storms and hurricanes. Some profited from this, others lost their assets. In other words, everything went along as usual. But now an event without equivalent is registered in the Bermuda Triangle. In 1840 a French vessel was discovered in the vicinity of the Bahamas, without a single soul on board, drifting under full sail.

Could it be that pirates were involved? But no, a valuable cargo was found aboard, including a large quantity of vintage wine. What self-respecting pirate would ignore such quality wine? Perhaps the vessel encountered a mighty storm, and the crew was washed overboard by towering waves? But this premise also had to be discarded—upon inspection, the vessel proved intact. Everything on deck, in the cabins, and in the holds was in its place, only the crew was missing.

Similar events began to multiply. Where do people vanish, under what circumstances do they abandon these ships, leaving their wealth behind? To this day, this remains an enigma, while among sailors the conviction that tragedies occurring in the Triangle are due to secret causes is only strengthened.

Rumors circulate about some sort of supernatural power operating in the region, triggering madness; due to a strong, unaccountable fear, sailors begin to dash about the deck, finally throwing themselves overboard and drowning. People not inclined to mysticism attempted to explain these Bermudan tragedies with strong ocean currents, underwater mountain ridges, and treacherous storms. But then, with the advent of aviation, planes also began to vanish…

After the Second World War, ocean and air travel between Europe and America increased—by sea alone there are nearly 150 thousand voyages per year. Is this the reason for the increased number of catastrophes in the Triangle?

After the war, traffic increased on other air and sea routes of the planet as well; however, nowhere did vessels and planes disappear as often as in the Northwest Atlantic. It's reported that in a 30-year period over 100 ships and planes vanished without a trace, an average of 4 disappearances per year!

On January 30, 1948, the 4-engine British South American Airways (BSAA) plane *Star Tiger* disappeared while in flight. With 31 passengers on board, it flew out from the Azores toward Bermuda. The pilot reported good weather and favorable flight conditions, but suddenly went silent. Forever!

On January 17, 1949, there vanished (under the same favorable weather conditions) another passenger plane, *Star Ariel*, with 20 passengers on board. 35 minutes after takeoff (the plane was flying from the Bermudas to Jamaica) radio contact with the plane was lost.

The plane was never seen again. Two days later, a fishing vessel disappeared in the same region. And in December of 1949 near the coast of Florida, 9 airplanes disappeared without any tidings! No distress signals ever came from them.

Sometimes Coast Guard dispatchers managed to intercept messages from someone suffering a calamity. But what they heard over the radio was so unusual: "compass needles are spinning unnaturally"; "the sky looks yellow"; "the sea looks strange, somehow."

Well, how can anyone gauge the level of danger from such intelligence? And among the US Coast Guard service there arose, willy-nilly, doubts about the psychic well-being of those sending such messages. Meanwhile, ships and planes continued to disappear.

The oddities of the Triangle don't end there. Here, in clear weather, people observe shining objects of spherical or elliptical form, which pursue ships and planes, gaining colossal speed. Sometimes, these light effects take the form of "pillars of fire" that seem to rise up from underwater and come to rest against the clouds. Many eyewitnesses relate that, upon the emergence of these light apparitions, ships and planes lose radio contact, and navigational instruments fail.

Another oddity—the absence of any trace of these catastrophes. Yes, in contradiction to the purveyors of mysteries, in the waters of the Triangle, there sometimes are found fragments of ships and planes, wrecked boats, life vests, etc. But other events, when even after the most thorough search, no trace of the disaster can be found, boggle the imagination.

We may also relate how planes flying across the Bermuda Triangle suddenly disappear from Coast Guard radar screens. Pilots and sailors become disoriented here. Something confusing happens to time as well—every so often it seems to slow down. This list of mysterious phenomena in the Northwest Atlantic may be extended, but later on we will look at every riddle separately.

The Bermuda Triangle acquired its sinister fame not so much from the number of catastrophes occurring there, as from precisely those mysterious phenomena accompanying them. I'll make a qualification here: analogous phenomena have been observed in other regions of the Earth's sphere, though nowhere as often as in the Bermuda Triangle. For this reason, I will refer to them as **Bermudan phenomena**, no matter in which region they occur.

In one of the reviews dedicated to the Bermudan enigma, there is expressed the conviction that the riddle "will be solved in the near future: scientists have seriously undertaken it, equipped with the mighty technology of the 20th century."

These optimistic words were written in 1976, a year after the introduction of a proposal in the General Assembly of the UN to create a special commission to study the mysterious catastrophes and disappearances in the Bermuda Triangle. Many years have passed since then...

1.3. Sensation of the Age

When it became clear that continued search for Flight 19 was senseless, a commission was organized to investigate the incident. Various accounts began to enter its purview. Here is one of them. Not long before, Navy command dispatched ships and planes to search for the crew of Flight 19, pilot Lieutenant P. Vershing was conducting a training flight near the Florida coast. As he returned to base, his plane mysteriously deviated from its course by 76 kilometers.

When ground crew later inspected the navigational equipment of his plane, everything functioned normally. Vershing sent an appropriate report to his command, but due to rapidly developing events, when all branches of the military onshore were scrambling to respond to the alarm, the incident was simply forgotten. But now Lt. Vershing's report crossed the desk of the investigative commission.

It's very possible that the disruption of compass function on the planes of both Vershing and Flight 19 may have been due to the same cause, which may have been strong magnetic turbulence arising in the area. Vershing was not overtaken by the tragic fate of the Flight 19 pilots, evidently because the coastline of Florida was within his line of sight. This suggestion would also explain the radio static that marked attempts to contact the disoriented group.

It's well known that between Florida and the Bahamas lies a region in which radio contact is almost always impeded, as surmised, by magnetic ores lying at the bottom of the ocean. But complications during flight arose before disruption of radio contact with Flight 19. Therefore, the cause is not a magnetic anomaly…

Let's allow that, having lost their orientation, the pilots, instead of flying west toward shore, headed instead over open ocean and, when their fuel ran out, made a forced water landing. In this case, they may have been saved: the planes were equipped with life rafts, which in a catastrophic situation would automatically drop into the water and inflate.

Those few minutes when the plane is still floating on the surface would provide ample time for people to scramble into them. Perhaps this is how it happened, but after all, not one of the 14 members of Flight 19 managed to survive. Let's say that waves washed them off the rafts. But then, they also had life vests…Rafts don't drown. Five planes—five rafts. The fact that after a thorough search not one of them was found puts into doubt the possibility of a landing on water.

Among the reports received by the commission there were those that pointed to the possibility of an explosion. A member of the crew of a freight plane that happened to be flying over the Northwest Atlantic on the ill-fated day saw a "pillar of light" that may have been taken for flames. If this were an explosion, then it was extraordinarily powerful, because according to the witness the column of fire rose high above the water.

Into the commission's purview came yet another report. At 19 hours, 50 minutes the crew of the ship *Gaines Mills* saw in the sky a "ball of fire" in exactly the same place where, according to calculations, Flight 19 then found itself.

Let's postulate that this also was an explosion. But it's impossible that five planes could blow up simultaneously! Could it be that the Martin Mariner flying boat exploded? Nothing of the sort! As became clear, radio contact with it was maintained even after the appearance of the light effect. If, however, there was no explosion, then what was the nature of the "ball of fire" and "pillar of light"?

Incidentally, the simultaneous explosion of the five planes of Flight 19 cannot be completely discounted. Among the commission members was a representative of the secret service. He knew that at one time Reichsfürer SS Himmler was investigating an unusual method of destroying enemy planes.

The method was made possible by the fact that flying planes emit a great amount of fine carbon dust in their wake. These carbon "clouds" would then ignited by a rocket. The person proposing this idea, German professor Hans Erhardt, was at that time confined in an internment camp. He made contact with American intelligence and told about his invention. The professor reported that his method was successfully employed in air battles over Europe.

Although Germany had capitulated several months earlier, the possibility cannot be excluded that there remained unconquered fascist groups. But how could these have turned up on the very shores of the USA? Clearly, from submarines. From the start of 1942, German submarines lurked in the Northwest Atlantic, spying on American transport ships carrying armaments and technical supplies to their European allies.

The Germans carried out warlike maneuvers basically at night; during the day they sunbathed, swam in the ocean, and even organized sporting contests on nearby uninhabited islands. Possibly, the crew of one or even several of these German submarines found themselves a haven on one of these islands, and the above-described ignition of carbon clouds may explain a simultaneous bursting into flames of the five planes.

Another form of sabotage is possible. It's well known that among the armaments of the Soviet Army were F-10 radio mines that were detonated at a distance by radio signal. In the fall of 1942, German sappers managed to discover one, but it took them over a year to copy the Soviet model. We may surmise that, not long before their deployment, German saboteurs planted explosive devices into the engines of the American planes, and that the signal was sent from a German submarine.

However, this scenario is strongly confounded by one circumstance. Flight 19 originated at Fort Lauderdale and flew toward Miami, while the Martin Mariner search plane flew from the airbase at Banana River. Is it possible that an organized net of saboteurs was active along the entire Florida coast? Not credible.

Besides, this version doesn't explain the other riddles. For example, the Martin Mariner search plane encountered a zone of strong turbulence at 1,800 meters. By all calculations, it should have already been in the same quadrant as Flight 19. But Lieutenant Taylor didn't report any turbulence. It looks like these powerful air streams were operating within a severely restricted area.

And further, in relatively calm weather, from whence came the big waves observed during the first hour of the search? And the enormous distances between them! Is there a connection between the disappearance of the planes and this phenomenon in the sea? And—most incredible: why did such meticulous searches fail to find not only fragments of the planes, but not even patches of oil, usually present on the water surface when a boat or plane has gone down?! Riddles, riddles…

The results of the investigation may be judged by the comments of commission members, which were subsequently publicized:

"Members of the commission are unable to put forward even a partway reasonable conjecture that might shed some light on this incident…"

"They vanished without a trace, as though they'd flown to Mars…"

"There has never been such a thing, that the military had to examine a case of the traceless vanishing of a group of bombers. Such an event in peacetime presents an unsolvable riddle."

In its time the disappearance of Flight 19 was spoken of as "the greatest mystery in the history of aviation." Precisely following this event, the Bermudan theme was elevated to the rank of "Sensation of the Age." But this is not the only reason I began the book with an examination of the tragedy of Flight 19. In fact, this incident contains in itself many of the same mysterious phenomena that I've called Bermudan phenomena. And the path toward solving the Bermudan riddle, finally, leads through an explanation of their nature.

1.4. In Search of a Hypothesis

1963 was named by the press a "prolific" year in relation to the Bermuda Triangle question—five disappearances were registered that year in the Northwest Atlantic. Precisely since that "prolific" year, news of perplexing events in the Bermuda Triangle began to be systematically reported in the international press.

But, according to German journalist Helmut Hüfling, what inspired absorption with the theme among the reading public was the emergence into the world of two books by American author Charles Berlitz. "Charles Berlitz," writes Hüfling, "presented the material in a more engaging manner, without aggravating complications, and thus evoked the interest of a wide circle of readers."

In Berlitz's books the reader finds not only gripping descriptions of events, but also becomes acquainted with other existing versions. If one casts a mental glance over them, it may be observed that they seem to split in two tendencies. The versions of the first tendency explain the various tragedies via a complex of particularities unfavorable for flights and sea voyages: tornadoes, hurricanes, swift currents, underwater volcanoes, treacherous reefs; versions of the second tendency explain Bermudan events via machinations by an unknown civilization. (Proponents of this second direction I will henceforth refer to as **civilizants.**)

Berlitz pays only glancing attention to versions of the first tendency, though he describes in detail and cultivates the idea of the invisible presence on Earth of representatives of an unknown civilization, and acquaints the reader with the authors of these versions. George Harder, for example, believes Earth to be a unique "cosmic zoo," in which guests from the cosmos conduct broad experiments in selection and genetic engineering.

John Spencer writes: "Since it's impossible to find a halfway plausible explanation for incidents of complete disappearances of huge ships on calm seas, as well as planes not long before their scheduled landing, I'm personally forced to think that they could only have been snatched from our planet."

Versions about the machinations of extraterrestrials base themselves on the frequent appearance of UFOs in the sky over the Bermuda Triangle. As though, displacing the space-time continuum, intelligent beings from distant worlds, or possibly from a parallel universe, somehow open a window onto circum-terrestrial space, and this window opens precisely onto the Bermuda Triangle.

Another version belongs to Ivan T. Sanderson. According to this version, cosmic guests who visited Earth several eons ago established technical complexes on the floor of the Atlantic Ocean, for the purpose of transmitting to their planet information about the development of humankind. This information is not transmitted continuously, but at specific intervals, with the aid of generators of massive energy. And if a plane or vessel finds itself in that region of the Atlantic when these generators are operating, catastrophe occurs.

All these versions, without much thought, could be attributed to the fantastical, were their authors people far removed from science. But Spencer, as noted by Berlitz, "is a member of an investigative group studying UFOs, which includes highly placed officials of the American government, overseeing the Navy and NASA."

Harder is a professor of engineering. Sanderson (also a professor) is the founder of an Association for the investigation of anomalous phenomena. It may be noted, of course, that versions of the second tendency do not become less fantastical for being expressed by scientists. However, we are not in a hurry—the problem is actually more complicated than it may seem at first glance. We will attempt to reason it out ourselves.

Any catastrophe requires a clarification of causes, and here people rely on accumulated experience. Thus, if the aviation disaster took place in favorable weather, the cause is attributed to "internal" factors—technical malfunction or pilot error. If, on the other hand, the disaster occurred at the time of a sudden hurricane, the cause is seen as a result of capricious elements. But let us pose a question: Can the complete disappearance of five bombers and one flying boat be explained by human error or technical malfunction?

You say this would be impossible. I agree. That means the cause is external. Then the next question follows: can the total absence of any traces of the catastrophe overtaking all six aircraft be explained by any natural cause? Now, this has given you pause. It may be that on December 5, 1945, in a limited stretch of the Atlantic, there simultaneously arose all the elements we're familiar with. The imagination paints a strange picture: a magnetic storm disrupted radio communication and functionality of compasses; a hurricane sheared off the wings of all six planes, and concurrently active underwater volcanoes sucked into their bowels all traces of the catastrophe. However, the reasonable mind balks at admitting the possibility of all these elements being in play at the same time.

If, however, in opposition to the reasonable mind (and sometimes it's necessary to think in opposition to it), we allow the possibility of such a concurrence, then the thought of the meddling of supernatural forces involuntarily arises. But since we are not inclined to mysticism, we'll have to discard the idea of an accidental, simultaneous convergence of various factors, and return to the idea of a single common cause. It's a vicious circle, isn't it? And only one way out of it presents itself: to abandon the attempt of attributing these events to blind elements, and address the cause, working reasonably and purposefully.

Since there's no basis for supposing that the Bermudan catastrophes, stretching over many years, are orchestrated by some sort of terrorist organization, the idea involuntarily presents itself that these are machinations of an unknown civilization, probably extraterrestrial. You are, of course, confused: the conclusion to which relentless logic has led us definitely smacks of the plot of some outdated fantasy novel. But there's no other way out, because to explain the simultaneous disappearance of six airplanes by other causes is completely impossible.

The idea expressed by Sanderson became the foundation upon which Berlitz built his own hypothesis of "pressure from below." I imagine that readers unfamiliar with Berlitz may be interested in becoming acquainted with it.

The hypothesis rested on two supporting points. The first was a publication about how, on the floor of the Bermuda Triangle between the Florida peninsula and the Bahamas, there was discovered a large elevation resembling a pyramid. This object was located not far from that region where, according to Atlantologists, there was once a huge island. This discovery on the ocean floor allowed Berlitz to suggest that this "pyramid" was built by denizens of the ancient Atlantis.

The idea is, naturally, controversial, particularly if one considers that the floor of the Northern Atlantic is littered with (mostly cone-shaped) volcanoes. Let's allow, however, that the pyramid was constructed by Atlantians. But this still leaves unexplained the disasters of the Bermuda Triangle—so what if the bottom of the ocean is a graveyard for the ruins of an ancient civilization! However, from the pyramid it's easier to construct a bridge to the second supporting point. The thought runs like this: if there exists a "manmade" pyramid, it's *highly likely* that there also exist other artifacts of the drowned civilization…

The second supporting point of the hypothesis is laser illumination from the bottom of the sea. Here Berlitz cites the observation of Edgar Casey (1877-1945), a famous American spiritualist and clairvoyant. Casey "clearly saw," at a depth of one and a half kilometers in the vicinity of the island of Bimini, sources of energy in the form of crystals, created by the inhabitants of the antique Atlantic.

As is well known, crystals serve as an important element in some lasers. Casey died 15 years before the appearance of the first lasers. This circumstance, in Berlitz's opinion, proves the clairvoyant capabilities of Casey, and consequently confirms his own hypothesis: the technical devices of the Atlantians have functioned on the ocean floor for thousands of years; the energy they radiate to this day influences compasses and electronic apparatus of vessels and aircraft traversing the Bermuda Triangle.

Berlitz devotes so many pages to Atlantis lore that a question involuntarily arises: what goal, exactly, is the author pursuing—to explain the riddles of the Triangle, or to strengthen his hypothesis about Atlantis? "The existence of Atlantis in the past and the existence today of the Bermuda Triangle," Berlitz writes, "present in themselves two riddles full of mystery, the answers to which are hidden in the depths of the Atlantic Ocean. The solution of one riddle will lead, evidently, to the solution of the other."

But as to which of these riddles Berlitz considered "first," and which "second," he remains silent. Meanwhile, in conducting our investigations it is imperative to clearly separate the factual material into that which needs to be explained, and that by which explanations need to be made. To base the previous existence of Atlantis on the events in the Bermuda Triangle is impossible. Because their causes have not yet been solved. But even the riddles of the Triangle are impossible to explain by information about the legendary Atlantis...[2]

Berlitz: "The majority of commentators regarding events in the Bermuda Triangle limit themselves by characterizing them as riddles defying solution. But some of the more thoughtful investigators consider that the mysterious disappearances of ships, planes, and people are connected to the activities of reasoning terrestrial and extraterrestrial forces—a point of view arising, perhaps, because there is no other logical explanation for these phenomena."

In this conclusion of Berlitz there lurks, if one looks closely, a logical error. Anecdotes are built on such errors, one of which I allow myself to present here.

In the course of some archaeological digs in Egypt a wire was discovered. Arab scientists concluded that during the time of the pharaohs there already existed a wire telegraph. An archaeological dig took place in Israel too, but no wire was discovered. This allowed the Israelis to announce that, during the time that the Egyptians made use of the wire telegraph, the Jews already had a wireless one! Truly, the conclusion is obvious: the absence of proof cannot be taken as proof.

Why is it that none of the existing versions has received popular acclaim? Why does the Bermudan question remain, over decades, the "riddle of the age"? There seem to be two reasons:

First, although versions of the first tendency do base themselves on relevant science, they are all essentially "tailored" for specific incidents. The adoption of a hypothesis (or theory), however, consists in the ability to explain an aggregate of similar phenomena, to discern among them a common pattern.

Second, although versions of the second tendency do encompass the full set of Bermudan riddles, they are all characterized by an absence of proof. Not every guess, meanwhile, qualifies as a scientific hypothesis, only those proposals affirmed by facts and human experience.

---

[2] "Lovers of romantic dreams about drowned Atlantis are prone to indulging in fantasies," wrote the noted traveler, Thor Heyerdahl. "But the best thing they can do meantime with respect to critically minded scientists is to say: you haven't yet disproved the existence of Atlantis." The "hypothesis" of Berlitz essentially represents an attempt to explain some riddles with other riddles. For such methods of proof science has yet to invent a name. I would call them collectively—"Atlanto-Bermudiana."

The inability over decades to come up with a feasible concept applying to the Bermudan issue as a whole has allowed some to cast doubt on the existence of the problem itself. "Seeking a common cause for all the disappearances in the Bermuda Triangle is no more logical than seeking a common cause for all the automobile accidents in Arizona," writes one of the investigators. He concludes: "The curtain over the mystery begins to part if we stop seeking a common theory and begin to investigate each incident separately."

When, about twenty years ago, I announced to my friend that I intended to write a book about the Bermuda Triangle, he, having a greater acquaintance with the issue began to try to talk me out of it. "The subject is obsolete," my friend said. "It's been ages since anything has been written about the Triangle."

That's because there haven't been any new ideas about it, I answered. "Mountains of books have been written about the Triangle," said others. "The theme has exhausted itself." Even more books have been written on the theme of love, I pointed out. But can anyone claim that the subject has been exhausted? It will never be exhausted, because love is an eternal mystery.

Of course, riddles of natural science, unlike the theme of love, can't exist forever. As regards understandable skepticism, I must note that in all time periods there have been people who regarded the theme of love, that most splendid of emotions, if not with skepticism then with relative equanimity. But does this cancel out the significance of the theme itself?

"The Bermuda Triangle is a tough nut to crack," my friends persisted. "People have tried to solve the riddle, whose professional expertise at least has some relevance to the subject: geologists, physicists, oceanographers. But you—you're a doctor. So tell us, what does your psychiatry have to do with the problem of the Bermuda Triangle?"

Honestly, at the time I myself didn't have confidence in success. But as someone famous once said: "The traveler will conquer the path!"

Clearly, daring alone isn't enough to construct a serious concept in the field of natural science. But I had a few prerequisites allowing me to take up this complicated task. They were hidden, oddly enough, precisely within the scope of my profession: learning in biology, physiology, and psychiatry will allow me, I thought, to solve this "great riddle of planet Earth." There was also, it's true, another prerequisite. But more about this in the next chapter.

1.5. "Life Has Gotten Boring..."

I'm writing this book primarily for those who are intrigued by the question: in what corner of the Bermudan riddle is hidden the key to its solution? This is why I now invite my readers to set off in search of this key. I warn you right away: hurry is not allowed. The experience of predecessors indicates that the journey awaiting us is among the hardest. They say that we learn from mistakes. This is true. But it's better to learn from the mistakes of others than from one's own.

The main danger awaiting us on this journey is an attempt, under any circumstance, to attach ourselves to some mystery. In general, the attraction to mysteries is a positive factor, stimulating the creative quest. But if this attraction becomes an end in itself, then the first goal may be lost. The danger is magnified in that it's often hard to discern the boundary beyond which the attraction to a mystery becomes an end in itself.

Sunken Atlantis Thrives And Kidnaps Ships And Planes!
Laser Beams From The Bottom Of The Sea!
Maybe You Can Find Your Friends In A Martian Zoo!

Similar colorful headlines dotted American newspapers at the start of the Bermuda Triangle boom. Many at the time were seized by panic, and prepared for the end of the world. Hypotheses explaining disappearances in the Triangle according to strictly natural causes somehow went unnoticed. People wanted to believe in mysteries, and that demand was fulfilled in plenty. In November of 1970, when a suspicion arose that the pleasure yacht *Jelly Bean* had gone down somewhere near the Bahamas, the newspapers immediately trumpeted the news.

Details were given about its passengers, its equipment, and the search process. But then *Jelly Bean* unexpectedly returned safely to its berth. The news of its arrival in Miami safe and sound didn't garner as much press as the initial news of its disappearance.

The pursuit of sensation sometimes brings blunders along with it. In 1980 the Soviet oceanographic vessel *Vityaz'* was heading back to its home country after a voyage in the Atlantic. During a short stop in Lisbon, a press conference was held aboard the ship. Questions were answered by Professor A. Aksyonov, leader of the scientific expedition. One of the journalists asked him: do you, as a professor, believe that once there truly existed an island that sank, carrying away with it a highly evolved civilization?

Aksyonov replied indirectly, noting only that "certain data from archaeological investigations present a definite interest." In this connection he cited underwater surveys in the Northern Atlantic that had been conducted by a marine expedition from Moscow University. On the way home, the professor was astounded to learn that the expedition to which he had referred…had discovered Atlantis. This news was broadcast by major Western wire services.

The ship's radio station received a radiogram for Aksyonov: Moscow was demanding an explanation. Other telegrams were received: the publishers of a range of newspapers and magazines were requesting the timely preparation of material about the sensational discovery.

Back home, numerous letters awaited Aksyonov, among them one from Charles Berlitz.

Berlitz wrote that beneath the ocean in the region of the Bermuda Triangle there exists a strange, 500-foot tall pyramid. This author of sensational international bestsellers was suggesting a collaboration with the Soviet scientist. "I will locate the pyramids," Berlitz wrote, "and prove that this is none other than the perished ancient civilization, and then all historical textbooks will have to be rewritten."

Aksyonov had no wish to participate in such a project, and declined the collaboration. "I've had enough of the mysteries of Atlantis," he declared. "And for this reason, please rid me of Bermudan mysteries as well. They don't exist."

Thus, the second danger lurks in excessive skepticism, which may lead to oversimplification, even to the complete denial of the issue itself.

The success of any enterprise, as everyone knows, depends in great part on the personal qualities of the "undertaker." I volunteered to proclaim the search for the key to the Bermudan riddle, and this obliges me to present to my sojourners if not proof, then at least a few credentials inspiring hope for success. Here I must confess that for a long time I didn't know anything about the Bermudan riddles. Of course, I read newspapers and magazines, eagerly sought in them unusual stories of all kinds.

The thing is, I was born and spent a major portion of my life in the USSR, where I could become familiar with news of the scientific world only through Soviet sources. And for many years these sources simply silenced the Bermudan subject. It was only in the '70s that news of the mysterious events of the Northwest Atlantic began to seep into the Soviet press. And they had to be somehow explained. But explained Soviet style, without frightening the peacefully laboring populace with tales of visitors from the cosmos! It was necessary to turn to scientific authorities for help.

The choice fell on Professor L. Brekhovskikh, leader of the Oceanographic commission of the USSR Academy of Sciences. In 1976 the Communist Party newspaper *Pravda* published an article by the eminent scientist, in which the "Bermudan myths" are subjected to crushing criticism. Professor Brekhovskikh states, in part, that in order to explain the Bermudan tragedies there is no need to turn to mysterious forces.

"The accident rate is heightened in this region due to more complicated hydrometeorological factors, arising from the influence of the warm waters of the Gulf Stream. Moreover, its relatively strong current carries away wreckage from aircraft and marine vessels suffering mishap. To a great extent this explains their disappearance without a trace."

There you go—simple and intelligible!

Soon after the publication of this article its author received a letter from a reader in the Moscow region: "Until your article came out, there was some mystery at least in the ocean," the woman complained. "You've destroyed it with your article. As a result, life on Earth has gotten very boring…"

Yes, grown-up people who have left childhood behind still want to believe in some kind of mystery, in something magic, suddenly become real. All those who never stop hoping, believing, and, most important, reaching the goals they aim for!

"The Bermuda Triangle does not contain within itself anything inaccessible to the understanding, nothing supernatural," Professor Brekhovskikh assures the Soviet reader. "What can account for the ever-multiplying scary stories about the Bermuda Triangle?" he asks.

And answers: "The appearance of myths…is connected to the pursuit of sensation, so characteristic of the organs of the press in capitalist countries, and their concurrent wars with each other. And it must be said that for many years the Bermuda Triangle has served them well. Wishing to amaze the reader, the authors of many articles find it easy to corrupt the truth, and often offer up deliberate fictions."

It was common to encounter similar comments in the Soviet press. They touched upon not only the issues of the Bermuda Triangle, but on other controversial themes such as ufology and parapsychology. Understandably, the writers of these critical articles—scientists and journalists—were expressing the opinion of Party ideologues. But is it possible to insist that they were always right? Familiarity with such criticism could not but inevitably be reflected in my own perception of "Western-style sensations." I was trained, you might say, to approach them cautiously.

On the other hand, my way of thinking was also formed by an inner protest against the "materialistic" world view imposed on Soviet citizens, with no room for mysteries. I, however, did believe in mysteries and—I'm not ashamed to admit it—believe in them to this day. It seems to me that a healthy skepticism in conjunction with a measured aspiration to partake of a mystery may serve as the best prerequisite for the success of the imminent journey. And so, onward!

1.6. A Man Named Kusche

Two paths lie before us. Already, at the start of each, we can see forks in the road. It makes the head spin to contemplate which way to begin. Hmmm…here's what I suggest. Let's start by selecting some particularly perplexing event in the history of the Bermuda Triangle. Let's see what other writers make of it, and try assigning a preference to one of them. And then, we'll see—the preliminary inferences we're able to make will show us the way forward.

An especially puzzling occurrence seems to me to be precisely the story of Flight 19: here we have not one, but simultaneously five disappearances! I decided to learn in more detail about "the greatest mystery in the history of world aviation." But the material with which I had to become acquainted immediately threw me into confusion.

There lives in the US, a man by the name of Lawrence Kusche, a former commercial aviation pilot, and subsequently an instructor at an aviation school. After leaving commercial flying he eventually went to work as a librarian at Arizona State University. In his spare time the former pilot searched through old newspapers and archival material for everything he could find about events in the Bermuda Triangle.

As a result of many years of painstaking research, he gathered some unique material which he synthesized in his book analyzing 53 incidents, selected by himself.[3] But, contrary to other authors, the former pilot didn't see any riddles in the Bermudan subject. He declared the events described by the *civilizants* to be mere speculation, and united them under one title—"Legend." Kusche's book was widely cited, especially by the foes of mysteries, and the author himself is recognized by them as practically the leading authority on the Bermuda Triangle.

---

[3] Lawrence David Kusche, *The Bermuda Triangle Mystery—Solved* (New York, 1975).

"In many cases the facts quoted in citations of official documents differ significantly from facts reproduced in the Legend," Kusche writes in the Foreword to his book. "Let the reader decide for himself which version more closely matches reality."

Well, let's follow this advice, let us compare the data from official documents, cited by Kusche, with the writings of Charles Berlitz. We'll focus, as before, on the example of Flight 19.

Kusche: "Many factors contributed to the loss of Flight 19, the most important of which was the failure of Lieutenant Taylor's compasses."

Berlitz: "The compasses failed on all five planes."

Kusche: "All the pilots, except Taylor, and all the crewmen, except one, were students in training."

Berlitz: "The members of Flight 19 were all experienced pilots."

Kusche: "Taylor was transferred to Fort Lauderdale shortly before this training flight, and was insufficiently acquainted with the region. As a result, he changed course many times, leading his squadron back and forth."

Berlitz: This fact is not mentioned.

Kusche: "Although the weather was 'fair' when the planes took off, it rapidly deteriorated. Search planes reported extreme turbulence and unsafe flying conditions."

Berlitz only indicates favorable weather at the moment of takeoff.

Kusche: "Although Taylor did allow a plane with a working compass to fly the lead position, he never turned over command of the flight."

Berlitz: "Taylor turned the command over to Captain Stivers."

Kusche: "The disappearance of the Martin Mariner search plane likewise presents no mystery. The plane took off at 19 hours, 27 minutes, and the explosion occurred…exactly where the plane should have been after 23 minutes of flight…"

Berlitz: "The Mariner took off and sank into oblivion in clear, sunny weather. The light effect, taken for an explosion, was observed three hours after the disappearance of the plane."

Kusche: "The Navy Board of Inquiry…listed 56 Opinions on the basis of the evidence of 14 days of investigation. As stated in Opinion 37, the airplanes of Flight 19 made a forced landing on water, in darkness…and Opinion 38 states that 'the sea was turbulent, which produced unfavorable conditions for a water landing.'"

Berlitz: "The Board of Inquiry was completely baffled."

So, extraterrestrial forces had nothing to do with this typically unfortunate incident? But how does Kusche explain the disappearances in the Triangle? He doesn't. The former pilot maintains that "there is no theory that would explain this mystery," because the mystery simply does not exist. He brings to this point the following arguments:

- "Disappearances occur in all regions of the ocean, and even on land. In the course of my research I found that, since 1850, between the New England states and Northern Europe, nearly 200 vessels disappeared or were abandoned by their crew."
- "Some losses that occurred elsewhere have been 'credited' to the Triangle…If all the locations of 'Bermuda Triangle incidents' were plotted on a globe it would be found that they had taken place in an area that included the Caribbean Sea, the Gulf of Mexico, and most of the North Atlantic Ocean."
- "Some of the lost vessels passed through the Bermuda Triangle but it is not known that they vanished there."
- "The Legend of the Bermuda Triangle is a manufactured mystery. It began because of careless research and was elaborated upon and perpetuated by writers who either purposely or unknowingly made use of misconceptions, faulty reasoning, and sensationalism. It was repeated so many times that it began to take on the aura of truth."

If we agree with the above statements, it turns out that the Bermuda Triangle is not at all unique, that catastrophes don't take place there any more often than in other regions of the planet, dangerous to navigation and flight. In that case, we've made a fuss for nothing, fooled the reader with Bermudan fables.

Not so fast, reader. I assure you that everything in the Bermudan question is not as clear and simple as Kusche attempts to represent it. There's much in it that is interesting, curious, strange. And do you know who ultimately persuaded me of this? You won't believe it—the Soviet press!

On January 5, 1983, the popular Soviet journal *Literary Gazette* published a major article, "Do mysteries exist in the Bermuda Triangle?" This time around, Bermudan riddles are not, as before, labeled "myths," but are identified as "unknown natural phenomena."

A remark by our old friend Professor Aksyonov is cited: "It's entirely possible that discoveries will be made in the ocean of the most unexpected phenomena, which will become the basis for re-examining long existing ideas."

If a few years ago Professor Brekhovskikh insisted there was nothing mysterious in the ocean, now we have a citation quoting the learned German professor, P. Dietrich: "We cannot exclude the possibility of the existence of 'mysteries' in the Bermuda Triangle." Brekhovskikh's name is not even mentioned (though, after all, he's a Soviet scientist), but the American Lawrence Kusche is subjected to open criticism: "An entirely serious problem of natural science he attributes to ordinary trifles…"

And further. "There is nothing supernatural there, but there are serious deviations from the norm…There are many naturally occurring perils here. The turbulent Gulf Stream, unexpected storms, sharp differentials in depth, sudden gigantic killer-waves, water spouts, strong atmospheric currents that can hurl a jet liner a thousand feet down."

Why these "deviations from the norm" are concentrated in such a constricted area, and what are their causes—so far, there's no answer. That's the only sense in which a mystery exists. But nevertheless, it's a real mystery.

It's totally obvious: if one of the leading Soviet journals has acknowledged, finally, the objective existence of a Bermudan problem, that means—mysteries do exist in the Triangle! That means, not everything in Berlitz's books is a bunch of mystification.

It means that errors lurk in Kusche's writings. Let's check. If we manage to find in his book incorrect observations and "tricky" argumentation, that can serve as an indicator that evidence concerning the Legend have a basis in reality. They are what will serve as the chief reference points in our journey through the Bermudan question.

1.7. Lights Over the Ocean

Eighteen years had gone by since the birth of a daring and glorious ambition in the mind of Christopher Columbus—to reach the Eastern shores of India, having circumnavigated Earth's sphere. And so, on August 3, 1492, a flotilla consisting of three caravels set out from the port city of Palos de la Frontera on the western coast of Spain.

The ambition did not arise suddenly. Even in his youth Columbus discovers in himself an extraordinary attraction to geography, and passionately surrenders himself to subjects having to do with seafaring—astronomy, geometry, navigation. Mastering this knowledge, the youth constantly seeks and finds confirmation for his surmise that the earth is a sphere composed of land and water, so that it's possible to circle it and come back to one's point of origin.

His conjectures are augmented by deep religious emotion, which gives him a tinge of superstition, albeit lofty and elevated: he sees himself as an instrument in the hands of God, chosen from all other people for the accomplishment of a great purpose. In Holy Scripture he finds, as it seems to him, a prophecy regarding his impending discovery.

When Columbus was implementing his daring ambition, Europe was just waking up after its long night of religious fanaticism. At this time much of the geographical knowledge gained by ancient sages still lay forgotten, and in Spain, the Inquisition reigned alongside the monarchs. It was exactly in this time that Columbus stood before a gathering of university professors and religious personages at the Dominican monastery of Saint Stephen in Salamanca, in order to lay out his views and solicit funds for his expedition.

Columbus's arguments demonstrating the spherical shape of the Earth were met with the following: "Where do you think you'll find the idiot who will believe that there are people living in the antipodes, whose footsteps lie opposite ours, who walk with their heels in the air, and heads on the ground?

"That there exists on earth a place where trees grow with their branches pointing down, and where the rain, hail, and snow fall upward? This fabrication about the roundness of the Earth," concluded the opponent, "has given rise to inventions about antipodes with their heels in the air, since these philosophers, once they've fallen into error, persist further in their absurdities, one serving as the basis for the next."

And another member of the commission, not actually objecting to the hypothesis concerning the spherical shape of the Earth, added that, "even if the ship does reach India, it still won't be able to return, since the rotundity of the Earth is like the incline of a hill up which it would be impossible to sail, even with the most favorable winds."

There were also comments to the effect that the Earth's sphere is so big, that a voyage around it would require years, and those who venture upon it would perish from hunger and thirst, since a ship wouldn't be able to hold the provisions and potable water necessary for such a long duration.

Meanwhile, Columbus had for years collected and studied antique maps, many of which, as historians tell us, represent a mixture of truth and error. In them knowledge gained in antiquity, and acquired in more recent times, sits side by side with folk fables and fanciful speculation. This priceless collection of Columbus's was especially augmented after his marriage, since his wife was the daughter of one of the most celebrated mariners of the time, the Italian Bartolomeo Monis de Palestrello.

The supreme preparedness of Columbus, assuredly, played its beneficial role in the expedition, since even bold sailors, who had more than once encountered the perils of the sea, considered this enterprise of Columbus's to be not only risky, but reckless. It wasn't by accident that, among the crew on the three caravels, there were many sailors who set out on this voyage into the unknown under duress, and were overcome with doubts and evil premonitions.

It wasn't by accident that, when the ships were sailing past the Canary Islands and a volcanic eruption occurred on the island of Tenerife, the sailors were terror-stricken. The spectacle of the huge mountain, belching flames and billowing smoke, they interpreted as a bad omen.

On the evening of September 15, Columbus himself observes a strange phenomenon: the compass needle no longer points north, but is displaced nearly 6 degrees to the southwest. And the further west the ships sail, the stronger the deviation. The sailors find out about this and decide that some sort of unholy power is leading the ships off course, not wishing them to reach their goal.

The situation intensifies, when toward morning the mariners become witness to an even more impressive and mysterious phenomenon: upon their course, between the sea and sky, there suddenly appears a gigantic pillar of light. This time the superstitious sailors, taking this as a warning from above, stage a rebellion and demand that the ships turn homeward. However, the captain remains undeterred, and pacifies the crew.

Because of the feeble wind, the ships moved forward very slowly, and it took them a month to reach the Sargasso Sea. At times they were completely becalmed, when the air was so still it was possible to read a book by candlelight on deck.

The monotonous days dragged by. Once, a watchman on the mast cried out that he saw land. The happiness of the sailors knew no bounds. Imagine their dismay when it turned out that what he'd actually seen was a mass of burly, floating seaweed. This "living carpet," slowing the progress of the ships even more, stretched all the way to the horizon, seemingly without end.

The sailors were especially troubled by the uncommon stillness of the sea. They took this for a sign that land was yet far off, since closer to shore the sea becomes more turbulent; moreover, they were nearing the end of their supply of potable water. On September 25, a Sunday, there suddenly rose a large wave on a windless sea.

Columbus, convinced that he operated under the sleepless gaze of God, who was guiding him on his great enterprise, notes in the ship's log that the agitation of the sea was sent by Providence for the appeasement of the sailors, and compares it to the miracle that saved the prophet Moses from the wrath of Pharaoh. Before his internal gaze, over and over there pass, like fragments of a majestic panorama, bits of Biblical scenes: with a wave of the Patriarch's hand, the waters of the Red Sea part; over the exposed bottom, throngs of Israelites move toward the birthplace of their ancestors...

And then a new "miracle"—coming on deck on the evening of October 11, Columbus notices in the distance, above the surface of the water, some sort of flickering flame. To ascertain that he isn't hallucinating, he points out the object to his assistant, then to another member of his command. And they all confirm that they see bursts of flame over the water.

And the "miracles" don't end there. "The sailors were cast into trepidation by the spectacle of a meteor, or as Columbus noted in his diary, pulsating flame, that seemed to be falling out of the sky into the sea four or five leagues from them." This version is from Washington Irving's biography of Columbus. The writer clarifies that "meteors are always observed in clear skies at these latitudes. In the transparent air of an exquisite night, when every star sparkles with the brightest light, they sometimes leave behind them a glowing trail…that can very well be reminiscent of flames."

It's well known that in his ship's log, Columbus recorded facts, and only facts. But in the notes made by him on September 15, 1492, there is no mention of any meteor, only information about a "flame" that over some period of time "seemed to be falling out of the sky into the sea."

Descriptions of optic phenomena, registered by Columbus in the Sargasso Sea, did not attract the attention of scientists. This allowed Irving to indulge in fantasies: "One or two more times they saw…flares, as though it were a torch on a fisherman's skiff, intermittently rising and lowering on the waves, or in the hands of a person on shore who, going from house to house, was now hidden, now visible…"

Well, Columbus's biographer may be forgiven for the buoyancy with which he so poetically described the mysterious light phenomena in the Northwest Atlantic. In the description of major geographical discoveries these details aren't all that important.

But the fact that Kusche casually cites the speculations of the novelist (he also writes of "meteorites" and "a torch in somebody's hand") is, to some degree, strange. Is it possible that he, having studied numerous publications about the Bermuda Triangle, could be unfamiliar with these optic phenomena that have been observed at various times in this region of the planet!

It's not only initiators of mysteries who speak of baffling phenomena in the Triangle. Let's turn, for example, to a serious scientific institution like the Czechoslovakian Academy of Sciences. In the encyclopedic dictionary published under its auspices we find: "Bermuda Triangle…famous to this day for unexplained disasters and disappearances of ships and planes. Participants in a joint Soviet-American research program in 1978 recorded numerous optic phenomena here, as well as disruptions of the Earth's magnetism."

Supporters of Kusche express regret for the fact that Berlitz and his ilk were able to fool the authors of this encyclopedia article. Thus, the Czech geologist Z. Kukal writes that "Disruptions in the behavior of the compass needle may be elicited by the influence of short-lived magnetic storms, which happen everywhere, including the Bermuda Triangle."

Still, how does Kusche explain the displacement of the compass needle aboard Columbus's ship? Very simply: there are regions on the Earth where the compass points not to magnetic North, but to the geographic North Pole (true North); at the time of the incident described by him, Columbus may have been in one of those regions. Yes, there definitely exist regions on the Earth where the compass needle "looks" toward true North, where the magnetic declination is zero.

Here, obviously, it would be good to explain what magnetic declination is. Many people think that the needle of a compass, from any point on Earth, points to the geographic North Pole. In reality this isn't so; usually it deviates from true North in the direction of the magnetic North pole. The observed angle between true North and the compass needle is called the magnetic declination.

The degree of magnetic declination varies in different regions of the globe, which has necessitated the creation of specialized geomagnetic maps. In order to make clear just how important a role magnetic declination plays in navigation, suffice it so say that if the navigator of a ship or plane fails to coordinate the readings shown by his navigational instruments with the geomagnetic map, he can go off course and even become completely disoriented.

Could it be that in September of 1492 Columbus was traversing such a region of the Atlantic? It can't be excluded. Generally, if one believes Kusche, there is no connection between Bermudan phenomena. But, reasoning logically, such a connection must exist, since a concentration of anomalous phenomena on a relatively small patch of the globe is hardly likely to be accidental.

1.8. "Blow, Winds, Blow…"

On March 4, 1918, the coal collier *Cyclops*, displacing 19,600 tons of water, with 309 people on board and a cargo of manganese ore, set out from the island of Barbados (West Indies). This ship, one of the largest in the US Navy fleet, was headed for Norfolk (Virginia), but never arrived there. Despite the mounting of a search, no trace of the vessel was discovered. *Cyclops* never even sent an SOS.

A suggestion was made that the *Cyclops* had been torpedoed, but a study of German archives after the war revealed that there had been no German submarines in that region of the Atlantic. Besides, the Germans customarily made radio announcements about the destruction of important enemy ships; however, in none of their radio bulletins was the *Cyclops* mentioned.

Representatives of the Navy pointed out that the weather hadn't been bad, or in any case, not bad enough to send to the bottom of the sea a large ship that hadn't even seen nine years of service. The captain of the *Cyclops*, George Worley, had spent 28 years in the Navy and had commandeered this vessel from its maiden voyage in 1910.

After a lengthy search, a spokesman for the Department of the Navy announced: "The disappearance of the *Cyclops* is one of the hardest of riddles to solve…Not one of the theories put forward explains with any degree of satisfaction, how and under what circumstances this ship vanished."

This is how Kusche presents the Legend of *Cyclops*. At the beginning of our journey we agreed that, in our search for the key to the Bermudan problem, we would consider the writings of this author. And, judging by the initial analysis, it looks like we're on the right path. I'm endlessly grateful to Lawrence Kusche: with his "unmasking" of the Legend he has given me more food for thought than both bestsellers of Charles Berlitz.

Also, because he has given us access to a unique compilation of official communications—in the protocols of military institutions, trade and insurance companies, as well as the periodical press.

Virginian Pilot

April 5, 1918

"Loss of American Coal Collier. Another Mystery of the Ocean."

"Washington, April 14. The *Cyclops*, a large US Navy coal collier (…) did not arrive at its designated port on the Atlantic seaboard, where it was expected on March 23.[4] As reported by the Department of the Navy, there has been no contact with the vessel (…) since March 4, and its fate elicits serious misgiving (…) The report from the Department contains the following: "(…) It is difficult to explain why the *Cyclops* is late in arriving at its designated port, as there has been no radio contact with her since her departure from port on one of the islands of the West Indies.

"Weather along the route the *Cyclops* was following was quite favorable…Of course, the *Cyclops* may have been sunk by an enemy raider or submarine, but there is no intelligence indicating the presence of any trace of the foe in that region of the ocean (…) One of the two engines on the *Cyclops* had been damaged, and she was moving at reduced speed, using only one engine. But even if both her main engines failed, she still had the capacity to send a message by radio. The search for the *Cyclops* continues."

Virginian Pilot

April 16, 1918

"Officials refuse to believe that an enormous coal collier, displacing 19,000 tons of water and carrying, in addition to its crew, 293 people, could disappear without a single trace. For this reason, search vessels were given the order to search every meter of the *Cyclops*'s route, and to approach every one of the numerous islands found in the zone of her itinerary.

"Representatives of the Navy openly acknowledged that not a single theory purporting to explain the disappearance of the *Cyclops* (…) stands up to fact checking (…)"

---

[4] Here and in the next chapter, ellipses in parentheses are Kusche's.

The cause of the catastrophe may have been ascertained, if the sunken vessel had been found. However, the disaster could have occurred on any segment of its designated route. Where to search? Representatives of the Navy can only spread their hands, but Kusche is deeply certain that the *Cyclops* went down off the shore near Norfolk (Virginia), almost at its designated port. He gets this assurance from the declaration of US Navy diver Dean Hawes, who in 1968, at a depth of 180 feet, 70 miles east of Norfolk, happened upon a sunken vessel. "When Hawes was subsequently shown a photograph of the *Cyclops*," Kusche writes, "he had no doubts that it was the *Cyclops* that lay on the bottom of the ocean."

"If the vessel spotted by the diver was indeed the *Cyclops*, one can imagine what a ruin it had become, having sat underwater for almost half a century! On the bottom of the Bermuda Triangle rest the rusted-through remains of many vessels. Is it even possible to identify these remains at a depth of 180 feet, where visibility is far from the best? We'll answer the question thus: in some cases, it's possible, if one conducts a focused investigation.

"Perhaps Hawes conducted just such an investigation? No such thing. In order to begin an investigation of the unknown ship," writes Kusche, "Hawes had to rise to the surface, where soon a strong wind carried his vessel a great distance from the place of discovery."

And so, the identity of the remnants of the sunken ship remains unknown, since they were never even investigated! Where did the diver get his assurance that they were namely those of the *Cyclops*?[5]

It's rare for a theory to be based on only one supporting point. If only some augmenting evidence could be found to strengthen Hawes's claim! "The news that Hawes had stumbled upon the sunken *Cyclops*," we further read in Kusche, "served as the prologue to another important event, enabling the discovery of the key to the riddle."

---

[5] In order to show how much known methods of identifying sunken ships differ from the methods recognized by Kusche, I will cite just one example. In March of 1911, the Australian vessel *Yongala* disappeared. In 1958, one George Conrad, in the course of an underwater search, raised from a ship, located at a depth of 35 meters, a reinforced safe. The English firm, Chubb and Son's Lock and Safe Company, received a photo of the safe, and the serial number stamped on its side. A week later came the conclusion: "Safe No. 49825 was prepared by us in May of 1903, per order from the firm Armstrong for the ship *Yongala*."

It's well known that the most frequent causes of shipwrecks are strong storms. "And I came to the conclusion," Kusche advances his theory, "that if, in those fateful days, there had indeed been a storm in the Norfolk area, then the sunken vessel found by Hawes could be the *Cyclops*."

A guess needs to be supported with facts, however, and Kusche telephones a request for the archives of weather bulletins for the Eastern seaboard of the USA. In the bulletins it says: "At the beginning of March, 1918, there were strong winds of 30 to 40 knots in the Norfolk area. On the 8th of March they had almost died down, but strengthened on the following morning." In a word, at the beginning of March the Norfolk area experienced windy weather.

Now Kusche undertakes to solve the brain teaser by means of distances, the probable speed of the ship (taking into account the malfunctioning of one of its engines), and even the speed of the Gulf Stream. The result of these mathematical calculations shows: the *Cyclops* should have reached Norfolk not on the 13th of March, as indicated by the newspaper, but on the 10th. Why the destroyer of mysteries needed this data, specifically, will become clear from the following.

Weather conditions as a possible cause of the tragedy were, of course, discussed by the experts. But every time the event was analyzed, such a possibility began to look so unlikely that it was finally dismissed. The Department of the Navy pointed out that the weather had been "not bad enough to send a large ship to the bottom of the sea." But what are mariners' opinions to a former pilot! "I was deeply convinced," Kusche writes in the course of proving his version, "that, contrary to the statements of the papers, the Navy, and the captains of ships on that day [March 10], a storm blew up in the Norfolk region."

Kusche complains about the fact that, in discussing the possible causes of the catastrophe, the investigative commission and the papers didn't count this storm, and finds an explanation for it: spring gales and storms are less noticeable on land than at sea. The foe of mysteries needs, however, not simply stormy weather, but a storm powerful enough to sink a ship.

And he paints a frightful picture that, in his opinion, must finally convince the reader of the truth of his theory: "...The north wind, encountering the Gulf Stream, raises waves of gigantic proportions; subsequently, it's possible to suggest that a storm came in from the north. The wind tears furiously into the mighty ocean current, turning it into a raging, seething torrent that has buried more than one ship."

Truly, a storm as ferocious as the one depicted by Kusche would have to sink more than one ship. Perhaps on March 10, 1918, a few more vessels experienced disasters in the vicinity of Norfolk? Or even one vessel? We find nothing along those lines in Kusche's book. Why did the *Cyclops* have to go to the bottom, a practically new vessel, commanded by a captain with 28 years of service?!

Let's turn our attention, however, to some other lines from Kusche's book. They will make clear how clumsily the author prods the facts to fit his theory. "…The wind blew from the southwest, gradually strengthening, and by 10 o'clock in the morning on March 10, its speed was 58 miles per hour. After noon the wind changed…Toward midnight the wind died down."

Now is it clear why the foe of mysteries needed the *Cyclops* to reach the port of Norfolk not on March 13, as indicated in the press, but on March 10? Why, because, precisely on the night of the 10th to the 11th of March, the weather in the Norfolk area became calmer. And remained calm over the next several days.

Kusche: "There were no focused searches for the vessel observed by Hawes…It may be hoped that in the near future Hawes' vessel will be found anew, and identified. It could possibly be the *Cyclops*."

From the time the diver Hawes discovered the unknown ship on the ocean floor near Norfolk, until the day of this writing, more than half a century has gone by. The riddle of the disappearance of the *Cyclops* remains a riddle.

Let's suppose, however, that Kusche had been unable to make the storm "compute." This, of course, would have seriously tarnished his theory, but would probably not have forced him to abandon the effort to think of a new one. For the destroyer of mysteries has prepared an invincible argument for just such occasions…

San Juan *Star*, Saturday
October 16, 1971

"The proprietor of the commercial vessel *El Carib*, which is already four days late in arriving from Baranquilla, (Colombia) to Santo Domingo, announced on Friday that, in his opinion, the vessel had been seized and diverted to Cuba. Diego Bordas, the owner of the Dominican shipping firm

Bordas Line (…) reported that he was trying to connect by phone with Havana, to find out something about the missing vessel…

"'If the boat had turned over, or started to sink,' said Bordas, 'it would have triggered a system that automatically sends out a disaster signal. Besides that, both inflatable life rafts are equipped with automatic radio beacons.'"

It wasn't by accident that I turned from the story about the *Cyclops* to the one about *El Carib*. The thing is, both stories contain the same riddles. The 542-foot coal collier *Cyclops* wasn't some tiny dinghy that would tip over before a sudden gust of wind. Big vessels, before they go down, usually agonize for hours. Why didn't the *Cyclops* send an SOS?

Was there an explosion on board? But then a great deal of debris would have floated on the ocean surface, indicating the place where the vessel sank. So, it wasn't an explosion. Since radio contact with the *Cyclops* broke off on the day of its departure from port in Barbados, on March 4, the supposition follows that it met with danger at the very start of its route. But what sort of danger might we speak of if, at the beginning of March, the weather in the region of the West Indies islands was favorable?

The misfortune of *El Carib* also occurred in good weather. The vessel and its two life rafts were equipped with automatic radio beacons. Why didn't the vessel, when it encountered danger, send out a disaster signal? Where ever could it have evaporated to?

Kusche: "I didn't find any reports of storms in the region of the Caribbean Sea at the time when the *El Carib* disappeared. However, this doesn't at all mean that the weather was good."

1.9. Pirates, to the Rescue!

However mysterious these vanishings of vessels without a trace in the Bermuda Triangle may seem, arguments from the opposers of mysteries, such as tardy or insufficiently focused searches, are almost impossible to refute. And how to explain incidents where vessels considered to be lost unexpectedly turn up, but minus people? Sometimes, even with life rafts on board.

The sailing ship *James Chester* was adrift in the North Atlantic when, on February 28, 1855, it happened to be observed by the sailors of the vessel *Marathon*. When the two sailing ships drew closer together, the captain of the *Marathon* ordered a boat to be lowered and set out himself for the abandoned ship. No crew was on it. At the same time, all the life rafts remained in place.

In the holds and in the galley, there were enough provisions and potable water. The cargo was fully preserved. The rigging was likewise in good condition. Over the next few months, news of the fate of the command, or the appearance of someone from the crew of the *James Chester*, was awaited in all the ports. In vain…

In 1955, the half-submerged vessel *Connemara* was discovered in the Bermuda Triangle. Not a single soul was on board, though all the life rafts remained in place.

Of course, I didn't borrow all similar incidents from Lawrence Kusche. If the foe of mysteries did speak of crewless ships, he avoided the question of life rafts.

Virginian Pilot
February 5, 1921
"Yesterday evening it became certain that the schooner abandoned by its crew (…) is the *Carroll A. Deering* (…) In September of last year the schooner departed for South America under the command of captain Merritt, one of its owners; its other owner was J. J. Deering, who had named the vessel for his

son, Carroll. A few days into the voyage captain Merritt fell ill and was forced to return home (…)

"66-year-old captain Wormell, an old sea wolf who had retired from the sea three years before, nevertheless took command of the schooner…He successfully completed the trip to South America and back, as far as Diamond Shoals, where some sort of disaster overtook the *Carroll A. Deering*. But why the vessel, under full sail and without a single sign of damage, turned out to have been abandoned by its crew—that still remains a mystery. We have no news regarding Captain Wormell or the members of his crew…"

Riddles give birth to new riddles. At first there rose some speculation of a mutiny aboard the ship. But shortly afterward reports began to come out about how, in the region of the ocean, other ships had vanished. Subsequent incidents, naturally, were tied to the riddle of the *Carroll A. Deering*. A new version was born.

New York *Times*
June 21, 1921
"…The crew of the American vessel has disappeared without a trace, and we have reasons to believe that the sailors were forced under force of arms to board another ship and were either killed, or were taken to some unknown port.

"Of the other missing American vessel there has long been no word, and two other American ships have vanished under conditions allowing a connection of their disappearance with the history of the capture of the first vessel. The government of the United States is taking every measure to solve these riddles of the ocean…Government representatives don't deny that it's hard to believe that in our days, in the territorial waters of the United States, pirates are able to operate and, moreover, that the facts are such that this possibility cannot be excluded" (…)

The governmental department (…) The Department of the Treasury, and the Coast Guard (…) the Department of the Navy (…) and the Justice Department (…) consider that all these mysterious events are links in a single chain.

"A few months ago, the five-mast schooner *Carroll A. Deering* **out of Norfolk, Virginia**, was discovered in the region of the shallows called Diamond Shoals, off North Carolina, under full sail but without any crew, who had vanished (…) Judging by everything, the schooner had been abandoned in haste, and at a moment when it was found to be in good shape, with a large store of edible provisions. Apparently, the catastrophe occurred not long before the crew was ready to sit down to dinner…"

"Sometime later, on the shore near the place where the schooner had been seen, a bottle was found with a note inside (…) in which it was possible to read the following: 'We were taken aboard either a tanker or a submarine, and put in handcuffs. Pass this message on to company officials as soon as possible.'

"The crew of the *Deering*, together with the captain, consisted of twelve people. All have disappeared without a trace (…) Besides this, there disappeared not long ago (…) the steamship *Hewitt*, out of Portland, Maine" (…)

"A representative of the Department of Commerce announced today that two more American steamships have vanished under circumstances which (…) allow a supposition that they did not sink, but became the prey of pirates (…) These ships were not named and, regarding details of their disappearance, official statements are more than opaque."

New York *Times*
June 22, 1921
"Today the Commerce Department announced the names of three other vessels that disappeared near the Atlantic seaboard of the United States, under mysterious circumstances (…) which allow a connection of their disappearance with the abduction of the American schooner *Carroll A. Deering* (…) It is highly remarkable that all vessels disappeared at approximately one and the same time, and that there remained no trace of their destruction (…) Ordinarily, when ships vanish, life rafts, debris, or dead bodies are found at the place of catastrophe, but in the given situation, the ships and their crews vanished without a trace…

"The government (…) has requested American consulates in various parts of the globe to watch closely, whether any members of the crew of the *Carroll A. Deering* turn up there."

"Today the staff of the Weather Service advanced a theory that connects the mysterious disappearance in the North Atlantic of a dozen or more vessels with severe storms…"

The newspaper article of June 24, 1921, served as food for new speculations about the fate of the *Deering*. Possibly, guessed some, a strong squall overtook the schooner, and the crew, realizing the danger they were in, lowered the life rafts in panic and attempted to reach shore. Hard to believe, contradicted others, that the 66-year-old Captain Wormell, an old sea wolf, was so frightened by a strong gust of wind that he ordered his crew to abandon the schooner, which was left under full sail and without any injury.

Still, how to dispel the mystification? Could be, by developing the version about pirates? For that, though, more convincing arguments are needed than the aforementioned guesses. After all, a lot of questions are found in the "pirate" version, for which it's impossible to provide reasonable answers. Let's assume that the ships really were seized by pirates. But a ship isn't a needle. Why didn't one of them turn up later at some port?

Why, even many years later, wasn't there found some member of one of the crews: all told, they must number no fewer than a hundred people! Why, after abducting over a dozen vessels, did the pirates disdain the *Deering*, a vessel without a single injury? Finally, why, having captured the crew of the *Deering*, didn't they take with them the large store of provisions? It's clear that the American government had little faith in the search for pirates, and was examining this possibility only because it lacked a more plausible version.

A representative of the insurance firm, Lloyd's, "ridiculed the note in the bottle, and said that, going by his many years of experience, messages of that sort almost always turn out finally to be utter nonsense." And truly, the note in the bottle turned out to be a crude fake.

Kusche: "Although the note found in the bottle turned out to be a mystification, the question inadvertently arises: couldn't pirates have actually seized the vessel?"

Well, it's a familiar method of proof.

Times **(London)**

July 11, 1969

"Jones Smith announces that Donald Crowhurst, participant in a single-handed, round-the-world yacht race, seemingly perished yesterday evening, a few days from the finish. His trimaran, the *Teignmouth Electron*, on which he was supposed to win 5,000 pounds as the prize in the round-the-world *Sunday Times* Golden Globe Race, was discovered 700 miles southwest of the Azores.

"The owner of the yacht disappeared, and it was impossible to figure out what had happened to him. The watch journal of Mister Crowhurst, along with books, documents, films, and magnetic tapes, all remained in their places (…) An inflatable boat and a life raft remained aboard the trimaran (…) Crowhurst was the main candidate for the 5,000-pound prize for winning this nonstop round-the-world yacht race (…)"

So, an inflatable boat and life raft remained aboard the trimaran. A puzzling story? Not at all. Because there was announced a version about suicide. "Since there turned out to be no chronometer aboard the yacht," Kusche begins to speculate, "it's possible to suppose that Crowhurst approached the rail and, glancing at his watch, threw himself in the ocean at the time he had designated himself."

Why, though, did the English sportsman decide to leave life in such an extravagant manner? Kusche suggests, due to the gnawing of his conscience—"The thought of a cheater's win, evidently, proved unbearable to Crowhurst…" The basis for this assumption was provided by the suspicion of a correspondent for the *Sunday Times* that Crowhurst was radioing false information about his location.

We won't debate, to what extent the absence of a clock on the yacht may serve as evidence of suicide. Let's turn to other facts. At this time four other unmanned vessels were reported in the Atlantic. You note: the Atlantic Ocean is big, in what region were these vessels observed? In Kusche we read: "They were found in the same region of the Atlantic as Crowhurst's yacht."

Maybe there was a storm? No, "as investigations showed, all these yachts were discovered in fair weather conditions," Kusche confesses. "There were no storms occurring here, much less vessels meeting with disasters."

You ask, in what period of time were these vessels found. I reply: four vessels were discovered one after another between July 1 and July 8. Here is how Kusche presents these events:

***Times* (London)**
July 12, 1969

"As reported from aboard the motor boat *Mailbank,* on July 1, a 60-foot vessel was observed off the northwest coast of Africa, floating bottom up."

***Times* (London)**
July 12, 1969

"On July 4, a yacht on autopilot was observed in the central Atlantic, holding a course to the east. There was no one in the cockpit of the 35-foot yacht. However, as stated by a representative of Lloyd's, this by no means signifies that it was abandoned by its crew."

***Times* (London)**
July 13, 1969

"…This was the 20-foot yacht *Vagabond,* on which Peter Wallin departed from Stockholm solo, bound for Australia."

Info Journal
Autumn 1969

"On July 8, 1969, the English tanker *Herson* passed by an upturned yacht of unspecified dimensions…"

"Five abandoned vessels"—that is what Kusche named the chapter in which he recounts the story of Crowhurst. Here, though, it would behoove us to speak of six vessels, since a disaster befell one other participant in the round-the-world yacht race for the prize offered by the *Sunday Times.* This was Nigel Tetley, who was running ahead of Crowhurst. Kusche mentions this tragedy only in passing. Here are the sentences: "Believing the false reports from Crowhurst, Tetley decided that Crowhurst was catching up to him, and cultivated such speed that the yacht couldn't sustain it, fell apart, and sank near the Azores."

These are, of course, clumsy speculations, since racing yachts don't fall apart at high speeds, high speeds are what they're intended for. If, however, the yacht was disabled for some reason, then why didn't Tetley save himself in the inflatable boat, which is part of the equipment provided for yachts participating in round-the-world races? One thing is unquestionable: in the region where Crowhurst's trimaran was discovered, one other person disappeared without a trace.

Judging from newspaper articles, at a minimum one of the four yachts, *Vagabond*, was investigated. Certainly, the question of the inflatable life boat, on which Peter Wallin could have abandoned the yacht, was central to the conduct of the investigation. The journalists couldn't ignore this detail either— probably the question was discussed in the papers. Why does Lawrence Kusche pass over this vital detail in silence? Because, I expect, on the yacht *Vagabond* the inflatable boat remained in its place. Just like on the trimaran *Teignmouth Electron*.

"Trying to find one common cause for all the disappearances in the Bermuda Triangle is no more logical than searching for one common cause of all the automobile accidents in Arizona…" Remember these lines? They were written by Lawrence Kusche. But if we were to follow his recommendation to "commence the investigation of each incident separately," what can be said about the four unmanned vessels?

Nothing out of the ordinary, the usual disasters at sea. It's worthwhile, however, to take in these events as a whole, with one glance, since the following implications leak out on their own:

1. Kusche, enumerating incidents covered by the press, from one to four, eliminates from the order of events the disappearance of Crowhurst. If the incident with Crowhurst were to be plugged into the row of disappearances as the fifth point (his yacht was discovered on July 10), then the version about suicide begins to look even more doubtful;

2. Kusche draws the reader's attention to the various dates on which the vessels were discovered, thereby deflecting attention from the thought that the people aboard them, possibly, disappeared simultaneously, or proximately in time;

3. Kusche focuses his attention on the fact that Crowhurst's yacht was found somewhat east of the Bermuda Triangle. In reality, this circumstance doesn't make the incident any less mysterious: an uncontrolled yacht could have been swept from the scene of the incident by fast currents;

4. The coincidental occurrence in time and location of incidents in which there are discovered similar circumstances is evidence that they, these incidents, clearly arise from the same cause.

The question is, where did the people go? Until such time as there is an unambiguous answer to that question, all hypotheses have a right to existence, including the possibility of kidnapping carried out by passengers of UFOs, and a displacement of the time-space continuum. So say the *civilizants*. And it's hard to contradict.

1.10. We Can Only Guess

Miami *Herald*, Thursday
August 29, 1963
Kurt Ludtke and Larry Miller

"On Wednesday, two gigantic Air Force tankers from Air Force Base Homestead disappeared over the Atlantic Ocean with 11 aviators on board. A search plane from the Bermuda Islands reported that on Wednesday a large oil slick was observed on the ocean surface, and oceangoing vessels pulled out several life vests. So far there is no data indicating that these findings are in any way connected to the vanished aircraft…"

Miami *Herald*, Friday
August 30, 1963

"Following inspection of fragments of wreckage retrieved from the waters of the Atlantic Ocean, serious fears have arisen that the two vanished KC-135 air tankers collided in midair. Searchers have almost lost hope of rescuing anyone of the 11 members of both crews. Three life rafts and a flight helmet with the name of one of the missing aviators written on it were found in the ocean…At the Pentagon, representatives of the US Navy suggested that the planes collided as they were returning to Homestead after having completed a routine, though secret, task of refueling over the ocean…Only a collision could explain the sudden, simultaneous loss of radio contact with the two gigantic 'flying gas stations'…"

Miami *Herald*, Saturday
August 31, 1963

"On Friday, planes searching for the two missing jet tankers…came upon another set of wreckage, located 160 miles from the first. Vessels were unable to collect any fragments from the water before nightfall…The large distance between these regions lessens the probability of the theory that the catastrophe was the result of a collision between the two planes in midair. So far, the Air Force is not identifying a cause that could have led to the destruction of both KC-135s…"

Miami *Herald*, Sunday
September 1, 1963

"Another, similar region was observed on Friday, 160 miles from the first; however, on Saturday, search team members reported not finding any objects from on board a KC-135.'There isn't anything there except some seaweed, tree trunks, and an old buoy,' said Major Fred Brent, an officer of the rescue service at the airbase in Orlando."

From a report regarding the search for two KC-135 aircraft, 5th Rescue Squadron, 1—31.8.63.MAC:

"In the course of massive search efforts…in the region of (point "S")…numerous objects were found…Examination showed that they belonged to the two KC-135 aircraft. A life belt, LPU-2/P with the number 314, had been located on board KC-135, craft 38, and a flight helmet with the name "Gardner" belonged to one of the crew members of KC-135, craft 41. Two radio beacon outfits taken from the water belonged to the KC-135 planes, crafts 38 and 41."

It's possible that, in the materials gathered by Kusche, there are other evaluations by officials in the US Air Force regarding each separate thing. But Kusche cites only the abovementioned statements, in many places interrupting them with ellipses. However, an analysis of even that material which the former pilot has presented for our perusal allows us to argue with his conclusions and, on certain points, with the findings of the investigative committee as well.

Contrary to the opinion of the Pentagon, I suggest that the simultaneous loss of radio contact with the two planes may be explained not only by a collision in midair. Basis: the search for the two KC-135 planes coincided with the search for a Coast Guard cutter *Gila*—radio contact with the cutter and the planes had been cut off almost simultaneously.

Chronicle of the week

Wednesday, August 28

Two KC-135 aircraft experience a catastrophe. An oil slick on the surface of the ocean is observed from the air, and that evening search vessels pull from the water several life vests. On them there are no identifying marks.

Thursday, August 29

Another area of accumulated debris is observed, 160 miles to the southwest of the first. The significant distance between the two areas forces a rejection of the initial theory: collision in midair.

It's surprising—why were the fragments observed by the pilots of search planes not removed from the water as soon as they were discovered? Let's pause at this question. The correspondent for the Miami *Herald* (in the dispatch dated August 31) refers to the coming of darkness. But this explanation is more than dubious. Strange, indeed: a major search operation is underway, already for three days vessels have been plowing the ocean expanse, on Thursday alone the number of search vessels reaches 36.

At last, on Friday evening, another collection of debris is found in the water, possibly fragments from the second plane. They need to be immediately retrieved, since in this part of the ocean strong prevailing currents can carry away all traces of the catastrophe to the north. Could the searchers have possibly not known this?

Thus, we don't know the true reason interfering, before sunset on Friday, with the retrieval from the water of "suspicious" fragments, and their identification. But even this unusual circumstance points to the fact that, in the incident under investigation, all is not clear and simple.

Let's return to our chronicle.

Saturday, August 31

In the morning it turns out that, in the place where some sort of objects had been observed yesterday, now there is only floating seaweed, tree trunks, and a buoy. Instead, fragments from the second plane unexpectedly turn up in the same area where on Thursday fragments of the first plane were found, that is, where on Thursday and Friday they didn't yet exist.

Let's look at two variations of the occurrence. First variation: let's suppose that on Friday, fragments from the second plane were retrieved. Can the demise of two planes be explained by a midair collision, when the accumulated fragments of each one are separated by 160 miles?

Generally speaking, a distance of 160 miles does not in itself exclude the possibility of a midair collision, since accumulated piles of debris could be carried far away from each other by winds and ocean currents. The challenge, however, turns out to be more complicated than first supposed—after all, we don't know the direction of the winds in the northern Atlantic on the day of the incident, or even have proof that the fragments of the second plane were caught up by ocean currents.

Second variation: fragments of both planes were found in the same quadrant—point S. Does this testify to a midair collision? The answer would seem to be affirmative. However, the indication of a point on a map of the Atlantic (a circle, a square)—is in reality some area of ocean surface, measured with a concrete number of square meters. Or, maybe, tens of kilometers. Since we don't know at exactly what "point" the catastrophe occurred, such a space on a map can be designated at will. Do we know what area on the map is actually indicated by point S?

No, we don't. Taking into consideration that fragments from the second plane were found at point S two days after the observation here of fragments form the first plane, one arrives at the conjecture that the initial piles of debris could initially have been separated from each other by a distance of 160 miles, and then drawn closer together due to winds and ocean currents.

As we can see, the solution of the second variant is also ambiguous; we get different answers, and determining which one of them is correct doesn't seem possible. The difficulty of the problem is determined, apparently, by the fact that we are saddled with an absolute condition: the two planes collided in midair.

At the same time, Kusche invokes the report from the commander of the rescue squadron. But how?! "Although, from the statement of the Air Force, they should have maintained constant radio contact and not flown too close to each other, it appears that there occurred a collision."

Such logic could drive anyone out of equilibrium. Nevertheless, let us gather our patience and see how Kusche resolves this contradiction: "The speed of their [the two KC-135 aircraft] flight was so great, that the distance of a mile could be reduced to zero in some few seconds. And if they set out on a course that would lead to collision, the pilots wouldn't have time for evasive maneuvers anyway…"

Excuse me, but in the cited materials there's no mention of "one mile"—no one saw how close to each other the two planes were flying. Kusche directs the reader's attention to the question of **how** a collision may have occurred, but there's not one word regarding the more significant **why**. Perhaps there was a storm? No, as stated by a representative of the Schilling airbase, "the refueling took place in favorable meteorological conditions."

Finally, let's consider the question: what is the likelihood of the collision in midair of two planes if they are flying:

- in favorable weather conditions
- in the same direction
- visible to each other
- not too close together
- with pilots maintaining reciprocal radio contact?

As we can see, the incident acquires an uncanny mysteriousness: two planes suffered a disaster simultaneously, but in different places. It's hard to believe, of course, that such a coincidence could be caused by different reasons. But—devil take it!—what common reason could there possibly be? Yes, there seems to be reason enough to spread one's hands.

1.11. Fateful Coincidences

And so, the two KC-135 planes, like the planes of Flight 19, were flying relatively close together. In both situations, radio contact was broken off with all aircraft simultaneously. What then, did all five planes of Flight 19 collide in midair? Such a thing can't be.

Weather conditions? But the KC-135 catastrophe occurred in mild summer weather. This means that the cause of the Flight 19 tragedy wasn't necessarily a storm. This means that both KC-135s could have perished because of something other than a collision in midair.

Let's suppose that a compass on a plane goes out of order. What is the cause? You'll say the cause, most likely, is in the compass itself, in its mechanism. Well, but if two compasses simultaneously stop functioning? You reply that such a thing may also happen. However, let's take the next step. Let's suppose the compasses on five planes fail simultaneously…

"But why make such a supposition," you protest, "if it's known that the compasses failed only on Taylor's plane."

"Allow me to ask, how is it known?"

"What do you mean, how? That's what Kusche says."

"But Berlitz says something else: the compasses failed on all five planes. Why do we have to give more credence to Kusche?"

"Why, because Kusche bases himself on official documents…"

Kusche considers that the key to understanding what happened on December 5, 1945 in the northwest Atlantic should be sought in the transcriptions of radio conversations with the pilots. Some of these transcriptions he presented in his book. Let us attempt to analyze these events, turning to the same transmissions.

From the testimony of Lieutenant Robert S. Cox, senior pilot instructor at the airbase on Fort Lauderdale:

"Approximately at 15 hours, 40 minutes, making a flight around the airport, I heard some radio transmissions between vessels or planes. On frequency 4805 kHz someone was talking to a Powers. He addressed his interlocuter by that name, and never gave his call signal. A few times he asked Powers about the displays on his compass, and finally said: 'I don't know where we are. Apparently, we got lost after the last turn…' I transmitted by radio: 'This is FT-74, addressing the plane or vessel calling Powers. Transmit your call signal, so we can help you.' He said that his call signal was MT-28.[6]

"I asked: 'MT-28, this is FT-74, what happened?' MT-28 replied: 'Both my compasses went out, and I'm trying to locate Fort Lauderdale, Florida…'"

At first the dispatchers of the command post on shore thought just this: the compasses failed only on Lieutenant Taylor's plane.

16 hours, 28 minutes

Port Everglades to Taylor:

"We suggest giving command of the flight to the pilot on whose plane the compasses are in working order. He'll lead you to the mainland."

Taylor: "Roger."

Kusche insists: though Taylor gave the lead of Flight 19 to Captain Stivers, he didn't turn over the command to him. Why does the former pilot stress this seemingly negligible detail? Because this detail serves as an important element in the scope of his argument. Among the factors impeding the rescue of Flight 19, Kusche points to military discipline, "which deprived the pilots of the possibility of making independent decisions."

Along with this, he's forced to make a qualification: "although many of them understood that they were following a wrong course." According to Kusche, it turns out: the compasses on the plane of Captain Stivers functioned normally, and he would have led the squadron to the mainland, if not for the erroneous orders of Taylor.

---

[6] Taylor misspoke, the call signal of his plane was actually FT-28.

From the report of the investigating commission:

Fact 18.

Stiver, Powers, Gerber, and Bossi were Navy pilots who were completing a course in mastery of the TBM. **[Translator's note: TBM is the designation for a torpedo bomber built by General Motors.]**

Fact 19.

Eight of nine crew members were also completing a course in mastery of the TBM.

Leaning on Points 18 and 19 in the report from the investigating commission, Kusche attempts to present the matter as though, with the exception of Taylor, all the crew of Flight 19 were novices just learning the art of aviation and knowing nothing of navigation. These "essentially, trainees" apparently couldn't be trusted. Kusche is silent about the fact that the pilots of Flight 19 had already flown between 345 and 375 hours, including bombers.

On that ill-fated day the military pilots were conducting a routine training flight, a task as familiar as regular training sessions are for athletes. The crew members of Flight 19 were by no means sheep who follow their leader over the cliff. Yes, Taylor made the decisions. But, as follows from the transcriptions of radio conversations, he consulted with the other pilots.

So, what happened to the compasses? Kusche answers this question thus: "As every navigator knows, in a time of turbulence, or in a storm, the compass needle constantly fluctuates…Any pilot who once in his life has rocked the wings of his plane, any sailor who once in his life has encountered a storm, knows that it doesn't take anything to make the compass needle start to spin around its axis."

Aha, there was a storm. But, after all—logic!—in conditions of turbulence, the needles of the compasses had to fluctuate on all the planes of Flight 19. How can the former pilot insist that the compasses failed only on Taylor's plane?

And another doubt may arise in us—why, even in strong turbulence, don't military pilots hurry helter-skelter to return to their bases, but continue to carry out their task? Because on board the plane, besides the magnetic compass, they have been provided with a radio compass, which remains totally unaffected by turbulence.

In the technical literature it shows that the chief advantage of the radio compass "is the possibility of sufficiently accurate direction finding…in any conditions of weather and visibility." This navigational instrument automatically shows the plane the direction toward the receiving radio station. And the radio station is situated on the home airbase.

Something happened with Taylor's compasses, the command on shore decided. But it couldn't be that the radio compasses malfunctioned on all the planes. The pilots must have forgotten to engage them.

16 hours, 56 minutes

Port Everglades to pilots of Flight 19:
"Engage your radio compass…"

From the testimony given by Lieutenant Robert S. Cox:
"Then I radioed FT-28 again: 'Your signal is getting weaker. It looks like something is out of whack…' At that moment my transmitter (with automatic tuning)…went out, and I could no longer maintain the radio connection…"
16 hours, 45 minutes

Port Everglades to Dinner Key:
"Can you fix a course on FT-28?"

Dinner Key:
"Negative…We can't catch a signal from his identification system."[7]
16 hours, 47 minutes

Fact 27.
"The network of high-frequency compass locators along the Gulf coast and the eastern seaboard of the US received an order to get a position reading on all transmissions from FT-28."
16 hours, 49 minutes

---

[7] An opinion was expressed that the pilots, conserving their fuel, were flying at a low elevation; that was why they weren't picked up on radar. But transcriptions of radio transmissions indicate that FT-28 was flying at a height between 3,500 and 4,500 feet.

Port Everglades to Dinner Key:

"Did you get a reading on FT-28?"

Dinner Key:

"No."

Port Everglades:

"Request permission to activate network of compass locators along the entire coast to find lost squadron."

Dinner Key:

"Affirmative."

Rough draft accident log:

"Connection with FT-28 intermittent. Port Everglades unable to transmit a single message to FT-28. Fort Lauderdale can hear radio conversations between the planes."

Taylor to Port Everglades:

"I can barely hear you. My reception is growing weaker…"
17 hours, 50 minutes

Fort Lauderdale to Port Everglades:

"All rescue stations on the coast are given the order to seek FT-28 and his squadron. All bases south of Banana River are put on number one alert."

At last there was success in fixing a position on Flight 19. The given coordinates had to be immediately transmitted by cable [by teletype] to the coast and search service stations.

Conclusion 21.

"At a critical point the teletype…of the air and sea service south of Banana River went out of order."
18 hours, 24 minutes

Port Everglades to Banana River (via telephone):

"All stations have to try to establish contact with FT-28 on frequency 4805 kHz...The situation is serious, as the planes lost orientation at 18 hours, 22 minutes, and by 19 hours, 30 minutes their fuel will run out. System TVPL (teletype) to the north of us is out of order. Also, hold the "Dumbos" (search planes) on your base in readiness."

From the testimony of Captain 3rd rank Claude Newman, commander at Port Everglades:

Question: "Did Port Everglades maintain radio contact with any of the lost TBMs at 18 hours, 22 minutes, when you got their coordinates?"

Newman: "No, sir, we didn't have two-way contact for nearly an hour."

From the evidence given by Captain 3rd rank Richard Baxter, US Coast Guard, 7th Marine district, Miami:

Question: "Which of these (Marine) bases were capable or had the possibility of transmitting the fixed coordinates to the TBM squadron?"

Baxter: "As far as I know, none of these stations was able at that time of day to operate adequately..."

The thought occurs that the problems with radio connection and compass failure on the aircraft of Flight 19 were caused by some common reason. Let's see, however, how the event unfolds further.

From the evidence of Lieutenant Samuel M. Heinz, duty radio operator at the dispatch command center on the Naval air base at Fort Lauderdale:

"From 16 hours, 10 minutes to 19 hours, 04 minutes, the lost planes were sometimes heard loud and clear, and at other times were not audible. Sometimes the radio station at Port Everglades picked up transmissions that weren't heard at the dispatch command center in Fort Lauderdale, and vice versa...The command dispatch center (Fort Lauderdale) wasn't able to get confirmation of even one of the messages sent from the planes of Flight 19."

The lost squadron may have been saved if Taylor had been given his coordinates.

Conclusion 22.

"Not a single operation of the marine rescue services attempting to help Flight 19 was able to establish a connection with it after the determination of its coordinates."

The lost squadron would have been rescued, had the search planes received its coordinates.

Conclusion 23.

"In not one single operation involved in aiding Flight 19 by delivering to its planes the coordinates fixed at 17 hours, 50 minutes, were special transmissions or broadcasting programs utilized."

Even more, another possibility was never utilized. One of the "Dumbo" search planes took off at 18 hours, 20 minutes; its pilot knew Taylor's coordinates. However, he was unable not only to see the squadron, but even to contact Taylor by radio.

Kusche: "...His antenna quickly iced over, and he couldn't maintain radio contact."

The icing over of an antenna can indeed worsen radio contact. But there are typically several antennas installed on a plane, tuned to different frequencies. Besides, the pilot of the "Dumbo" plane wasn't the only search participant who by that time had received the coordinates of Flight 19.

At 18 hours, 45 minutes, a group of planes from Vero Beach, and later from Daytona Beach and Banana River, headed for the quadrant in which Taylor had gotten lost. And all of them maintained radio contact with each other and with command on shore. Why didn't the antennas on these planes ice over?

1.12. Disappearance of the Martin Mariner

As we recall, the Martin Mariner hydroplane that flew from the air base at Banana River with a rescue team aboard, headed for the region where Flight 19 had experienced distress. It too never returned to base. Traces of the catastrophe were not found.

If, in a given region, five planes vanish without a trace, and after that—one more, then it's highly probable that both incidents share the same cause. So logic insinuates. But, following this logic, the investigating commission would have faced an unsolvable problem. This brainteaser is quickly resolved, however, if the two disappearances are examined independently of each other, assigning them different causes. This is exactly the course followed by the investigative commission.

Conclusion 37:

"The planes of Flight 19 made a forced water landing in darkness, in the region east of the Florida Peninsula after 19 hours, 04 minutes."

Conclusion 53:

"The training plane number 49 (Martin Mariner) exploded in the sea due to an unknown reason or reasons…"

Kusche attempts to guess the reason why the Mariner hydroplane might have blown up in midair: "The Mariners were nicknamed 'flying gas tanks' because they always contained a lot of gasoline vapor, and a surreptitiously smoked cigarette could at any moment cause an explosion."

Encountering similar commentary in Kusche's book, we must exercise especial vigilance. The former pilot, unmasking the Legend, appeals in certain instances to reports by investigative commissions. Such documents normally elicit trust. Here, however, it's imperative to consider that experts aren't always able to establish the cause of a catastrophe. And, as demonstrated, in such cases the more realistic version is put forward.

For instance, in the case of the two KC-135s, members of the commission suggested the more (in their opinion) likely cause—a collision in midair. But Kusche presents this suggestion as established fact. And now as well. Conclusion 53 states: "due to an unknown reason." But Kusche fantasizes about a smoked cigarette. Personally, I don't want to believe this version: if military pilots nicknamed this plane a "flying gas tank," they'd be hardly likely to smoke cigarettes in it, let alone surreptitiously.

From the testimony of Captain 3rd rank William J. Laurence, senior assistant to the commander, US Coast Guard, Marine airbase, Banana River:

"At 21 hours, 12 minutes, I got a message from the joint operations center in Miami that at 19 hours, 50 minutes, the ship *Gaines Mills* observed an explosion at a point with coordinates 28 degrees, 59 minutes North, 80 degrees, 25 minutes West. The explosion, as reported, was powerful, and the column of fire lasted several minutes. It may be surmised that this was the training plane Number 49, with which we tried unsuccessfully to establish contact..."

From the testimony of Captain 3rd rank William T. Murphy, US Coast Guard, Miami:

"We got the following message from the US Navy ship *Solomons*: "Yesterday evening we were able to observe a plane on our radar screen...At the same moment when the *Gaines Mills* saw flames, and in the very same place, that plane vanished from our radar screen and didn't reappear."

Citing the evidence from the Coast Guard officers, Kusche concludes: "The Mariner took off at 19 hours, 27 minutes, and the explosion occurred...exactly in the place where the plane should have been after 23 minutes of flight."

There is some basis for disagreeing with Kusche: in different authors describing this event, the time frame varies from five minutes to half an hour after takeoff. Meanwhile, the time factor appears to be central to understanding what happened to the hydroplane—if the light effect on the ocean was not an explosion, then the Mariner likely disappeared for the same reason as the planes of Flight 19.

There are two versions before us. To which should we give our preference? Yes, a bright flash of light was seen from on board the *Gaines Mills* at exactly the time when the radar on board the vessel *Solomons* registered the disappearance of the plane. In the very same section of sky. Here, time and space seem to have converged on one point; and if we examine events from this point (of view), then the version about the explosion of the Mariner will seem preferable. If, however, we bring into the circle of investigation some additional evidence, then that version will begin to seem vulnerable…

From the testimony of Captain 3rd rank William J. Laurence:

"…The captain of the steamship *Gaines Mills* stated that the plane burst into flames in midair, swiftly fell into the water, and exploded, and that members of the ship's crew saw an oil slick and fragments. The commander of the division of surface vessels, based at New Smyrna, reported subsequently that he didn't see any fragments…"

A bright flash in the sky, naturally, brings to mind the explosion of a plane. And when, in the darkness and above the surface of the water, there rose a fiery pillar, it may have seemed to some of the sailors that the fragments of the plane were burning. In addition, it behooves us to consider the weather: "total overcast, occasional heavy rainfall, wind…the sea was very turbulent."

In the circumstances indicated, there exist all the preconditions for giving rise to optical illusions. What they saw from aboard the *Gaines Mills* may have been, for example, a mass of seaweed, or simply trash thrown out of a closely passing vessel, that is, what is commonly called marine debris. Or maybe there was nothing there at all, and something only seemed to appear to the sailors.

On board the Mariner there was a great deal of lifesaving (unsinkable) equipment. Where did it all go? It's possible, of course, to bring in a favorite argument of the foes of Bermudan mysteries: in stormy conditions it took rescue vessels several hours to get to the area of the incident; during that time the remains of the hydroplane were carried far from the place where it fell by wind and ocean currents.

Let's allow, however, that, suspecting a plane crash, the captain of the *Gaines Mills* immediately set a course for the pillar of fire rising above the sea.

Would there have been traces of the catastrophe in that case? You suppose that, in any case, objects in the water would have been identified…

Exactly here, it's necessary to tell of one circumstance that (of course!) Kusche passes over in silence. The captain of the tanker actually did set course for the fiery pillar rising over the water, and reached that place in half an hour. But nothing, except for an oil slick, was observed here. Does an oil slick prove that a plane sank in that spot?

We'll give the answer to Kusche himself: "An isolated oil slick does not conceal any mysteries in itself, since on the ocean surface there constantly appear a great number of oil slicks of the most various origins." And further: "…We can't say with certainty where this oil slick came from, since no fragments were found in this area."

The cited statements refer to a different event, when the oil slick was found in a place totally different from the one wished for by the author.

We find another unusual circumstance in a book by David Groupe: "…The people on the tanker *Gaines Mills* watched the 30-meter high flames for a period of 10 minutes."

Question: could the burning fragments of the plane have caused a pillar of flame 30 meters high? In a very turbulent sea? For 10 minutes? This point could be argued, if it were not for one detail in the report of the investigative commission.

From the testimony of Lieutenant J. I. Bemmerling, pilot of training plane number 32, Navy air base, Banana River:

"…At approximately 21 hours, 45 minutes, we were given a mission to go into the region, situated about 25 miles to the east of New Smyrna, to examine the place of the explosion reported by the tanker…We commenced a broad, systematic search at the place of the explosion, investigating everything that appeared on the radar screen and the lights, observed visually; however, the search yielded no results."

Wait, what were these lights observed by the search planes? Nothing is said about this in Kusche's book. The destroyer of mysteries, having inserted many ellipses into the documents cited by him, forgot to drop evidence about lights over the ocean from the testimony of Lieutenant Bemmerling.

And so, on December 5, 1945, in the area of the incident, luminous effects of some sort were repeatedly observed. Allow me to make the following inference: the "flash" in the sky and the "fiery pillar" over the water's surface don't have a direct relation to the disappearance of the Martin Mariner hydroplane.

1.13. "Perhaps the Donkey is Yours..."

The folklore of central Asia features a well-known and remarkably colorful character: Molla Nasreddin. Stories about him are told to this very day. I'll give an example here. One day two villagers quarreled over who owned a donkey grazing in the meadow. Unable to agree, they asked Molla to help settle the dispute. He asked each of them for evidence.

"My donkey has a broken hoof on his left hind leg," declared one of the neighbors.

Molla examined the donkey's leg, ascertained the truth of the statement, and announced: "The donkey belongs to you." Then the other villager began to prove his claim.

"My donkey," he said, "has a big scar on its belly." They looked, and saw that it was so.

This time Molla confirmed: "Perhaps the donkey is yours." But here one of the gathered neighbors broke in.

"Molla," he said, "it's impossible for both sides in a dispute to be equally right."

Nasreddin thought a bit, and then solemnly concluded: "And you are also right, respected one…"

The more you familiarize yourself with literature about the Bermuda Triangle, the more convinced you become: Lawrence Kusche can't be believed on all points. Does this mean everything written by Charles Berlitz ought to be taken on faith? Not at all. Kusche justly accuses the authors of the Legend of consciously suppressing certain facts in order to lend a greater sense of mystery to events.

Berlitz, for example, doesn't mention that the angle of the fragments of the two KC-135 jets turned out to be closer to each other than originally supposed. But Kusche is guilty of doing the same, though with the opposite aim—to reduce the sense of mystery. Recounting the incident of the two KC-135 jets, he passes over in silence the simultaneous disappearance in the same region of the Coast Guard vessel "Gila," with which radio contact was also abruptly cut off.

Of course, the simultaneous cutting off of radio contact may have happened by chance. But this coincidence in the Triangle is not the only one.

Not for nothing a spokesman for the US Navy says: "One gets the impression it's as though the vessels and aircraft are covered by a gigantic electronic camouflage net. We know that something strange is going on there, we've always known about it, but haven't the faintest idea why this specific thing is happening. And we treat these facts with utmost seriousness."

Kusche writes: "In the Bermuda Triangle, any local magnetic disturbances are absent..." Meanwhile, Soviet geographer, Professor V. Kort, heading an investigative expedition in the waters of the Triangle, points out: "in this region there occasionally arise powerful magnetic disturbances, producing a danger of 'radio silence.'"

They say that the truth lies between two opposing viewpoints.

"In no way!" exclaims Goethe. "The problem lies between them." It was just such a problem that Molla Nasreddin encountered—in a dispute both sides can't be right. Although the evidence of the quarreling villagers was corroborated, the arguments they presented apparently turned out to be inadequate. Molla might have been able to reach an objective decision if he'd been able to identify additional circumstances, for example testimonial proof from the other neighbors.

We too must do something similar. When contradictions are found in the writings of Kusche and Berlitz, it becomes necessary to fill in the picture with testimony from other sources. Here we engage in a work similar in many ways to that of a physician making a diagnosis. If a patient complains, say, of headache, it's possible to suspect a migraine. But the cause of headache may also turn out to be high blood pressure.

The thing is, the headache is merely a symptom common to a variety of ailments presented in the clinic. It may be caused by a swelling of the brain, by migraine, or simply extreme fatigue. Knowing one symptom is frequently inadequate for making a true diagnosis. The physician must identify not one, but several symptoms to arrive at the syndrome that will allow an objective picture of the illness.

To repeat: I perceive the path to solving this problem lies in the shedding of light on the nature of Bermudan phenomena. After all, the catastrophes that occur in the Triangle are often preceded by all these mysterious manifestations—luminous effects over water and in the sky, interruption of radio contact, disruption of the functionality of navigational instruments, a loss of capability of orientation, and more.

And if it's impossible to establish the causes of many tragedies, it's not without reason to adopt something in the nature of a bypass maneuver, and begin by figuring out the nature of the Bermudan phenomena themselves. These, figuratively speaking, are the symptoms that need to be systematized and assembled into a syndrome.

A preliminary analysis of the event that occurred with Flight 19 allows us to think that we have "grabbed by the tail" the nature of two of the most commonly registered Bermudan phenomena: interruption of radio contact and deviation from course. It's true that the role of magnetic factors in Bermudan tragedies has long been affirmed by the *civilizants*. However, we'll continue our own investigation, and pull this thread out of the impossibly tangled knot of the Bermuda problem.

The published report of the commission investigating the disappearance of the flight 19 runs to 400 pages. On the basis of this document Kusche reaches a conclusion I consider necessary to quote in full:

"Many factors interfered in the rescue of Flight 19: malfunction of the compasses on Lieutenant Taylor's plane; one of Lieutenant Cox's radio channels going out of commission, which interfered with his ability to maintain contact with Lieutenant Taylor.

"The failure to put the lead plane under Cox's command; bad radio reception; a delay in the launching of rescue aircraft; the falling of darkness and a sharp deterioration in weather conditions; the impossibility of quickly establishing the position of Flight 19; the failure of efforts to transmit the coordinates of the squadron by radio, once they became known; the malfunction of the teletype cable; the icing of the antenna on the "Dumbo" search plane, flying out of Dinner Key; the military discipline that prevented the pilots from making independent decisions, though many of them understood that they were following an incorrect course; the circumstance that Flight 19 was last to make its training run that day.

"If even one of these factors were absent, the episode of Flight 19 may have ended completely differently. At least one (and maybe more) of these planes would have returned to their base, and the entire dramatic episode would have quickly been consigned to oblivion, instead of turning into 'the most mysterious event in the history of global aviation.'"

Amid the factors enumerated by Lawrence Kusche we may distinguish the "symptomatic":

- Malfunction of the compasses aboard Lieutenant Taylor's plane
- Failure of one of Lt. Cox's radio channels
- Poor radio reception
- Impossibility of quickly identifying the position the squadron
- Failure to transmit the coordinates of the squadron by radio
- Malfunction of the cable (teletype) connection
- Icing over (?) of the antenna on the search plane "Dumbo"

Even to the untrained eye it's apparent that all seven factors have something in common: they are **essentially technical**, and in their physical nature—**electromagnetic**.

Why did Kusche disperse these technical factors among other, non-technical ones? Why, because, if they were presented together, as we have just done, a "syndrome" immediately emerges, and doubt arises—is this coincidence of technical factors random? This doubt, in turn, gives rise to the thought of "a gigantic electronic camouflage net," covering the area of the incident of December 5, 1945.

1.14. Occam's Razor

One criterion of truth, it's well known, is praxis, or human experience. Namely, the knowledge obtained through experience (observation, experiment), confirms or disproves this or that hypothesis. It's probably for this reason that, when the conversation turns to the Bermuda Triangle, I'm sometimes asked whether I've ever been in that region myself. I answer, no, I have not. In itself, a visit to that region of the Atlantic would at best confirm the existence of certain puzzling phenomena, but it wouldn't produce an understanding of their nature.

There's an extensive amount of literature about the Bermuda Triangle, and no one is prevented from making use of the information presented there, in order to rework or rethink it. But how do we relate to publications whose authors totally reject the existence of a Bermudan issue? How do we ascertain key questions pertaining to the issue, when even sympathizers express contradictory opinions about them?[8]

---

[8] Here are a few examples of statements by opposers of Bermudan mysteries, considering the frequency of disasters in the Triangle:

L. Brekhovskikh, member of the Academy of Sciences, USSR: "Facts confirm that in the Bermuda Triangle...there is observed an elevated number of marine and air catastrophes."

Z. Kukal, Czech geologist: "Statistical data indicate that the number of vessels and planes that have 'vanished' in the Bermuda Triangle does not elevate the average index in other parts of the global ocean."

D. Repin, member of the Geographical Association of the USSR: "In the region of the Bermuda Triangle, catastrophes occur more often than in any other region on Earth."

G. Massi, American journalist: "Berlitz claims that the frequency of disappearances in the Triangle and other parts of the world simply doesn't bear comparison. However, no numbers confirming his position are provided, which fully conforms to the traditions of pseudoscience."

How can we seriously relate to books in which gaps in knowledge are filled with unproven guesses, or even outright fantasies aiming for effect? How is it possible, finally, in this ocean of literature, to distinguish facts from scientific-sounding speculations, puzzling events taking place in reality from fictions? It's clear that, relying solely on the help of evidence contained in the literature about the Bermuda Triangle, getting at the truth is impossible…

I will attempt to formulate the problem more precisely. A series of authors maintains that an elevated percentage of accidents to vessels and planes is observed in the Northwest Atlantic. But in publications about the Triangle I wasn't able to find results of statistical investigations. What seems more substantive, however, is the circumstance that some tragedies occur in good weather, while their causes remain unresolved. But is it possible to ascertain at least a common characteristic of these catastrophes?

Generally, yes. However, the conclusion, resulting out of a seemingly simple analysis, gives rise to bewilderment. Judge for yourselves. Logic: if these catastrophes occur in favorable weather, they must be caused by technical malfunctions on board, or by human error. But with respect to these events in the Bermuda Triangle, this logic doesn't apply. The thing is, catastrophes facilitated by internal causes occur, to a known extent, accidentally.

Therefore, due to statistical regularity, they are distributed among all marine and air routes of the planet equally, and don't influence the percentage ratio of catastrophes in separate regions. What, then, follows from this? Just that the elevated percentage of tragic incidents occurring in fair weather is the result of external causes. Causes invisible to us, and therefore incomprehensible. It's in this, actually, that the secret of the Bermuda Triangle is concealed.

In the array of external causes, a special place is occupied by magnetic disturbances and electrical fields. But what causes these physical phenomena? Secret underwater complexes of an unknown civilization? Or geophysical processes unknown to science, in which Earth's magnetism and atmospheric electricity play a role?

In the 14th century, the Franciscan monk Occam introduced a principle, vital to science, that can be reduced to a brief formula: when investigating some unknown phenomenon, it's best to avoid attribution to supernatural causes until such time as attempts to explain the phenomenon according to natural principles have been exhausted.[9]

In other words, the search for a key to the unknown should first be conducted within the framework of the known, without inventing new facts. This principle was called "Occam's Razor" because, right from the start, it cuts off attempts to explain a phenomenon with extremely complicated methods or with speculative versions.

Occam lived in the Middle Ages, when unknown natural phenomena, astounding the imagination, were explained through the intervention of supernatural powers. However, science today strictly adheres to Occam's principle. For example, regarding the issue of extraterrestrial civilizations, there is a principle known as "presumption of naturality."

It means, essentially, that every newly discovered source of cosmic radiation must primarily be considered natural, until such time as serious arguments can be put forth in favor of its artificial origin.

In step with Occam's principle, it behooves us also to consider versions dealing with natural causes of Bermudan tragedies. However, having embarked on that path, we immediately run into formidable contradictions. Let's take, for example, underwater volcanoes.

Their eruption really does produce gigantic waves, capable of swallowing any vessel in an instant. This is where we get the absence of SOS signals, or any traces of catastrophe. But why do planes vanish? Storms and hurricanes, of course, may be the reason for these disasters; as with vessels, so with planes. But how to explain catastrophes occurring in fair weather? Perhaps some other elements operate in the ocean? But these we don't know.

Here, it would seem, we may be allowed to switch to an investigation of the second tendency. But it turns out that the second tendency doesn't look any better. Let's allow that the disappearance of vessels is a machination of the underwater dwellers of Atlantis. But how do they manage (and what for?) to snatch planes out of the sky?

Or else: extraterrestrials abscond with vessels and planes in order to study the development of our technology. Then why are there occasionally found in the ocean all kinds of technical apparatus belonging to marine and air vessels

---

[9] William of Occam (or Ockham, 1280-1349) was a major figure of the Franciscan Order; he established the doctrine known as nominalism. From 1345, until his death, he oversaw the Franciscan Order.

that have experienced a catastrophe? Why do people vanish from certain vessels, while the equipment remains intact? It's possible to say, of course, that the purposes of extraterrestrials are impossible for us to understand. But in that case, no logic will help us get to the truth…

When you try to work all of this out, the scope of all the "hows" and "whys" invariably leads reason into a dead end. Then you lose patience and feel as though you're drowning in this Bermudan problem. Where to go further? A third tendency? However, if you, esteemed reader, attempt to come up with a third tendency, you can be certain that nothing worthwhile will come to mind.

Can it really be that the Bermudan problem is insoluble? Yes, we've taken up a difficult challenge, but backing off is not an option. We've already covered a good section of the way, and discovered something important: mysteries do exist in the Bermuda Triangle. Consequently, somewhere there has to lie the key to their resolution…

Not long ago, I recalled a game with blocks that I used to play as a child of kindergarten age. On the six sides of each block were depicted fragments of six different pictures. Turning the blocks this way and that, I could assemble the entire picture.

Six pictures were reproduced on the back of the cardboard box the game was packaged in (I only remember a dog and a flower in a pot), so it didn't take much effort, comparing the fragments on the sides of the blocks with the "exemplar," after some rummaging around, to get the desired picture. Something similar, esteemed reader, is required of us in the effort to construct a "General Theory of the Bermuda Triangle."

So, we have at our disposal a certain set of "cubes"—Bermuda phenomena. They must be joined together so that a harmonious natural-science picture is obtained. But you, I see, are slow, you are not sure that the mosaic folded by you will correspond to the natural "original". I can even understand you, I'm not sure either.

The fact is that by combining "puzzles" this way and that, we will get more or less significant compositions, but it is not possible to determine which of them is true—unlike a child's game, we do not have a "sample" for comparison. In addition, we are dealing not with six, but with a large number of Bermuda phenomena, the nature of which is largely incomprehensible.

The only way I see of solving this challenge is by systematic recourse to "primary sources"—works in discrete realms of the sciences, written by "narrow" specialists. Precisely in such literature, essentially textbooks, lies expertise, verified and confirmed by many generations of people. This reliable knowledge will, I hope, help shed light on the nature of Bermudan phenomena. The criterion for the veracity of the resulting natural-science picture must be the mutually dependent interaction of the "fragments," their conformity with known laws of nature…

Someone among my readers has probably just thought: the author has already guessed the nature of the light effects, he knows what is represented by the source of magnetic disturbances in the Triangle, and will now lead us in some new direction. Nothing of the sort. Things don't happen that fast. After all, besides magnetic disturbances and light effects, a whole range of other mysterious phenomena are registered in the Triangle, and as yet I have not the slightest idea how to approach them.

1.15. "And the Stink, Monsieur..."

In 2012 Russian media (citing American sources) published a hypothesis of Joseph Monahan and David May concerning the causes of catastrophes in the Bermuda Triangle. So as not to retell the gist of their hypothesis, Here are some excerpts from the publications.

"It seems that two Australian scientists, Professor Joseph Monahan and his student David May (Monash University, Melbourne), have uncovered the mystery of the Bermuda Triangle. The cause of mysterious disappearances of vessels, it turns out, is a natural gas—methane. Oceanographers investigating certain hazardous regions of the ocean bottom found areas of old eruptions with great concentrations of methane. According to the scientists' hypothesis, when methane is freed from naturally-occurring cracks in the ocean floor, it transforms into huge gas bubbles that subsequently expand diametrically, rise to the surface of the water, and burst there. Then the gas begins to rise through the atmosphere..."

"As demonstrated by a computer model, any ship that is caught in a mega bubble of methane instantly loses its flotation, sinking to the bottom of the ocean. These gigantic gas bubbles are also capable of knocking planes out of the air."

"Methane Bubbles Rising from the Ocean Bottom Sink Ships and Cause Combustion of Planes."

Here it should be noted that the similar hypothesis – about the death of ships – was expressed in my book "The Truth about the Bermuda Triangle" (Moscow, ed. AST, 2001). I believe that it will be interesting for the reader to trace the path along which I went to solve...

In 1959 the French geophysicist Haroun Tazieff participated in an expedition to Africa, with the goal of investigating the structures of major geologic rifts in the "dark continent."[10] These major rifts are elongated troughs in the Earth's crust, once filled with water, and forming the African Great Lakes.

The expedition came to Lake Kivu, on the shores of which lay the small Congolese town of Goma. At this time a group of Belgian scientists was there, who had discovered a large accumulation of combustible gases dissolved in the lake water—methane and hydrogen sulfide. For Congo, lacking in timber, and where gasoline obtained from the coast was very expensive, this discovery of practically free energy was a gift from the heavens.

Tazieff: "...*I met the scientists right at the moment when they'd just tested their system. They were smiling happily, they were in marvelous spirits, but as soon as the subject of results was mentioned, they immediately clammed up. The thing was, the head of their firm in far-off Brussels had compelled them to maintain utmost secrecy. It was only a lot later that I was able to get from them the details that interested me. In the meantime, in Goma, I was forced to extract information from Bruno, a pleasant African who served as captain on their investigative vessel. Bruno still couldn't get over what he had seen!*

"'They lowered a plastic pipe into the lake, it must have been 500 meters long,' he related. 'Then they started pumping. Then they disconnected the pump, but water kept gushing from the pipe all by itself. And the stink, monsieur, I can't even describe!'...

"The Congolese, as a rule, are fairly impervious to smells, so if Bruno was complaining about the smell of rotten eggs, there was no doubt—we had the characteristic sign of hydrogen sulfide!"

I'm not citing Tazieff because he was a well-known specialist in the field of volcanology who was not "siloed" within the parameters of his profession, or because his interests were broader—you might say, global. Or even because, though he was French, he happened to be fluent in my native Russian language. The thing, of course, lies somewhere else: in the work of this scientist I came across information that, as it seems to me, allows the simultaneous illumination of several Bermudan phenomena...

---

[10] Professor Haroun Tazieff (1914-1998) was the head of the Paris Institute's department of Geophysics, and Chair of the scientific council of the International Institute of Volcanology.

All forms of life, plant and animal, inhabiting the surface of the Earth and its waters, naturally come to the end of their existence and die, and their bodies, decomposing into their chemical elements, enter the eternal "cycle" of Nature. Those remains of marine organisms that the fish didn't manage to snack on settle on the ocean bottom and, decomposing, exude both hydrogen sulfide, and methane—that same gas we utilize in our homes.

Waters near the very bottom are also saturated with various gases that are the products of volcanism, which under great pressure escape through fractures and crevices on the ocean bottom.

Tazieff: *"In Lake Kivu…the lowest layer remained untouched by circulation: it's too thick, and lies under the weight of the top layer, which exerts on it a pressure of 25 – 30 kilograms per square centimeter. Over millennia, these bottom waters have been enriched with an unbelievable amount of gases. If the pressure were changed, the gases dissolved in the water would begin to rise to the top…"*

If the pressure were changed…In the consciousness—a "quick link," illuminating the memory of an array of Bermudan events. Lights! Lights over the surface of the water! This sudden flash of insight must be immediately verified. I sift (yet again) through the pile of magazine and newspaper articles from different years. Hurray, I found it! Information about the expedition of the Soviet oceanographic vessel *Vityaz* into the Bermuda Triangle.

From the article by Leonid Pochivalov, "Do mysteries exist in the Bermuda Triangle" (*Literaturnaya Gazeta* [*Literary Gazette*], 5.01.1983)
"…One evening the weather forecaster Yurate Mikolayunene anxiously informed us: 'The barometer is falling rapidly.' In a couple of hours, she alerted us with new information: "The barometer is falling precipitously.' The ocean was almost still, but the barometer column descended lower and lower, nearing dangerous limits…

"In her scientific journal the weather forecaster noted that on May 6 the falling of atmospheric pressure was unique in its rapidity—abrupt, like a mountain landslide. One of us couldn't sustain it; the ship's doctor determined that Suvorov had suffered a stroke. The seaman required immediate hospitalization. The nearest dry land—the Bermuda Islands."

Tazieff: "*...The unexpected source of energy is a reservoir with a width of 400 square kilometers and a thickness of 250 meters, that is, with a volume of 100 cubic kilometers...Besides, the source is practically inexhaustible, given that during the process of extracting the gas, more methane will be collecting in the lower layers.*"

The Atlantic Ocean is not some Lake Kivu, where gas bubbles up, and all it takes is to pump some water out of the depths. The ocean has a different magnitude. Here the pressure from upper layers of water on the lower ones may noticeably diminish during an acute descent of atmospheric pressure. This happens in the regions where cyclones and tropical hurricanes are born, something for which the northern Atlantic is particularly well known.

...Three days following the disappearance of Flight 19, Navy pilot Morrison was conducting a training flight along the coast of Florida. At 2 o'clock in the morning he saw lights below, and some sort of masses, resembling human figures in motion. I must be seeing things, thought Morrison. But when, having flown another 32 kilometers on a southerly course, he again saw lights on the ground, he radioed the information to his dispatch center. Could it be that the pilot was suffering hallucinations?

After all, the place of which he spoke was a total desert, filled with swamps stretching for many kilometers along the shoreline. Nevertheless, it was decided to check out Morrison's message. Another pilot, taking off for the indicated region, confirmed it: lights. In the morning the location was combed by military units, but nothing indicated that a few hours ago something had been burning here.

The proposition expressed by someone, that the planes of Flight 19, failing to reach Banana River, had possibly made a "swamp landing" and drowned, was rejected—after all, the lights in the night were seen three days after the disappearance of the planes. Most likely of all, it was methane burning, which, as is well known, forms in abundance in swamps. And the swamps of this region communicate with the ocean...

An indication of mysterious optical effects may also be found in Kusche's descriptions of the incident with the American military's Globemaster transport plane, which vanished in the Bermuda Triangle in March of 1951. Also, in the story of the disappearance (on November 9, 1956) of an American P-5M patrol bomber.

Journalist Michelmore, not at all a disdainer of Bermudan mysteries, cites in his sketch of the Triangle a story told by Jim Connors, a sea wolf who spent over 20 years on the captain's bridge.

"One time we were running at night, 150 miles to the southeast of Andros Island. There stood a complete calm, so I lay down to rest in my cabin. When the watchman raised an alarm to call me out to the bridge, at first I couldn't believe my eyes: the sea around me was burning. Not phosphorescing, but burning with real flames. I gave an order for full speed ahead—in seven minutes we burst through this 'marine bonfire.'

"I started to try to figure, what and how. The helmsman and the watchman claimed in unison that the water suddenly burst into flame all by itself around our *Coral Shark*, at the stern, and in front, and at the sides. If a pleasure yacht had been in our place, it would have been toast. Would you believe, a bonfire in the middle of the sea! It's a mystery, that's for sure…"

Later, in order to dispel the mystique, Connors expressed the conjecture that his ship may have possibly rammed through a barrel of gasoline in the dark. But such an explanation doesn't work for the incident with Columbus, when he saw lights over the ocean surface—500 years ago, gasoline wasn't manufactured.

Here it's more likely to be something else: if, at the time when combustible gas is excreted, the atmosphere is saturated with electricity, then all it would take is a spark for that particular place in the ocean to "ignite." Incidentally, traces of phosphine (a gaseous compound of phosphorous and hydrogen) contained in methane may self-ignite upon contact with air.

Only half a century ago it was considered that, with increasing depth, life in the oceans becomes increasingly impoverished, and the scarce inhabitants of the depths receive nothing but crumbs from the upper layers. However, subsequent investigations by oceanographers have forced us to re-examine the perception that the lower depths represent a "biological desert."

In some regions of the world's oceans the density of life in the deeps turned out to be even greater than in the upper layers. In one of his descents in a bathysphere Jaques-Yves Cousteau made the following note in his diary: "The deep-water layers of plankton reached such density, there was an impression that we were descending through a living purée."

If in various regions of the world's ocean the density of life differs, it follows that the concentration of combustible gases in deep water differs as well. But there probably doesn't exist in the world's ocean another place where the density of life could be greater than that of the Sargasso Sea. Here there's accumulated such a quantity of seaweed that it forms as it were a natural carpet, covering many square kilometers of the ocean surface. Columbus' description of the Sargasso Sea comes to mind – and one's hand involuntarily reaches for the reference books:

"The Sargasso Sea is located in the central portion of the northern Atlantic…"

"…nearly equivalent in area of territory to the United States of America; it stretches for more than 2,000 miles in length, and 1,000 miles in width."

"The sea got its name because of the huge quantity of sturdy seaweed floating on its surface, among which the Sargasso type stands out (from the Portuguese "sargassum", sturdy seaweed). Thanks to floating bulbs that resemble grapes, the seaweed stays on the water's surface…"

"The Sargasso Sea is inhabited by many unusual creatures…"

"There were rumors that ships' crews, finding themselves in this cursed sea, died an agonizing death by suffocation."

Indeed, in the literature about the Bermuda Triangle there is described a whole array of incidents when ships found themselves in whitish fog; at the same time, crew members developed dizziness, suffocation, and a strong tickle in the throat. Sometimes (if a sailing ship ended up in a zone without wind), death came from poisoning. From here came the ships with dead crew members.[11]

---

[11] Hydrogen sulfide forms in the presence of rotting organic matter, is also present in volcanic gases, and possesses strong toxic effects. In the presence of large concentrations (1 mg/liter and higher), poisoning develops instantly, manifesting in convulsions, swelling of the lungs, and loss of consciousness; death results from paralysis of the respiratory center.

1.16. How Ships "Disappear"

Among sailors there have long circulated rumors how in the waters of the North Atlantic ships sometimes plunge to the bottom, but so precipitously that it's impossible to say they "sink"—no, they simply plunge. And how this happens in places where white water appears. There have, of course, been skeptics who didn't believe these rumors.

But we can't disbelieve Columbus: twice the Admiral noted in his ship's log an abrupt change in the color of the water, where the sea suddenly became "white as milk, as though it had been mixed with flour." And there were later witnesses…

The boat was already underwater and, judging by the length of time Joe had been struggling to reach the surface, was at a depth of 20 – 25 meters. In those seconds that seemed like an eternity, Joe's consciousness was chilled by the thought that the *Caicos Trader* had probably shared the fate of his *Wild Goose*, and he would now become easy prey for sharks in the nighttime sea. Fortunately, the tugboat crew had noticed in time that the *Wild Goose* was being dragged underwater, and quickly severed the cable. Gulping salt water and already losing consciousness, Joe Tayley was hauled aboard.

"This incident," commentators remark, "serves as the best proof that vessels really do vanish under mysterious circumstances. The boat disappeared literally in a matter of minutes, leaving behind not a trace. Science is so far incapable of providing an explanation for such phenomena…"

On October 5, 1951, the vessel *Southern Isles* went down off the coast of Florida. This happened so fast the radio dispatcher didn't have time to send a distress signal. The captains of the vessels following behind the *Southern Isles* reported later that its lights simply vanished.

Tazieff: *In eastern Africa, to the north of the Great Rift Valley, there lies a little-investigated lake. It was discovered in 1888 by the Hungarian explorer*

Teleki, who gave it the name of the great duke, Rudolph. In 1886 Teleki had left Zanzibar at the head of a large expedition through desert regions, and came upon the southern tip of the long lake. There he happened upon an active volcano, named Teleki Volcano by the geologist von Kennel, a member of the expedition…In 1921 Mt. Teleki erupted anew, the ground shook, and a pillar of fire was reflected in the southern portion of Lake Rudolph…"

In the summer of 1966, in the region where Joe Tayley's vessel vanished, the tugboat *Good News* was pulling an empty barge from Puerto Rico to Fort Lauderdale on a 300-meter cable. At one point, captain Don Henry was urgently called on deck. What he saw defies description: on all sides, almost to its railings, the boat was surrounded by water that had suddenly acquired a milky color; even the surrounding air was somehow misty, and it was hard to tell where was water and where was sky, and where was the horizon dividing them. In this whitish fog the barge was invisible, and the thick tugboat cable became so taut it seemed about to burst.

For some reason, *Good News* began to noticeably lose speed. Probably the barge had been pulled underwater, and was dragging the tugboat behind it. Don Henry, however, didn't cut the cable, but instead gave the command for "full speed ahead." The engine whined, but contrary to expectation, *Good News* remained in place, and at one moment even seemed to be moving backward a little. At last, however, the tug managed to crawl out of the strange cloud; soon the barge showed up again, as well. The sea was calm as before, visibility was good. Later it was discovered that the 50 batteries powering the boat's instruments were completely drained, and all had to be tossed overboard.

Tazieff: *"…For hours we gazed at the water, unable to get our fill of looking. This was a pristine lake…About 20 kilometers in front of us the large, hilly South Island reigned over the waters. In the 1930s a group of American and English geologists were there. They landed on the island and without any incidents spent three days there, while the rest of the expedition investigated the northern part of the lake.*

"After finishing their work, the geologists got in their boat and…disappeared. In all likelihood, they were destroyed by a suddenly developing squall—one of those for which Lake Rudolph is famous…"

I haven't the shadow of a doubt: if Tazieff had become absorbed in the, seemingly far from his interests, riddles of the Bermuda Triangle, he would have been able to explain the true reason for the disappearance of the geologists. And the cause of one of the impressive Bermudan phenomena to boot—the disappearance of vessels without a trace.

It's even surprising that an investigator as astute as Tazieff didn't think of another possibility, only suggesting that the geologists' boat was overturned by a suddenly developing squall. But the volcanologist was so close to the solution!

Judging by everything, the north face of the volcano lines the southern part of the lake bottom, precisely where the bottom communicates with the volcano, and from where there occasionally erupt volcanic gases under tremendous pressure.

Powerful jolts on the ocean floor, elicited by volcanic earthquakes, may also cause the excretion of gas from the ocean depths. The likelihood of gas dissolved in bottom waters rising to the surface grows even greater if the underwater earthquake coincides with a sharp drop in atmospheric pressure, for example, in conjunction with a cyclone.[12]

As well, the density of ocean water (and its corresponding specific weight) over an expanse comprising tens, or maybe thousands, of square kilometers, falls acutely. It stands to reason that this would be reflected in the flotation of vessels...

In December of 1967, when strong northeast winds whipped the waters of the Atlantic into a "roiling, frothing surface," the 23-foot cabin cruiser *Witchcraft* disappeared off the shore of Florida. Dan Burack, the owner and an expert sailor, had taken his friend, the Catholic priest Patrick Horgan, on a "marine stroll" to admire the Christmas lights of Miami from the water. At 9 pm the Coast Guard received a radio message from Burack, saying the screw had struck something, and the engine had developed a strong vibration.

Though the boat lost its ability to move, they weren't in immediate danger; the hull wasn't damaged, and in any case the cruiser couldn't sink because it was outfitted with built-in flotation compartments, and all they needed was a tow into port.

---

[12] Acute fluctuations of atmospheric pressure may themselves appear as a "release mechanism" of an earthquake if, of course, the pressure in the Earth's core were critical.

After fruitless searches, a spokesman for the Coast Guard announced that "the cruiser and passengers have apparently vanished, and not perished in the sea." Such a statement wasn't made baselessly: if the disabled cruiser had become a toy for the waves, the prevailing northeast wind would have driven it to the Florida shore; even if a wave had overturned the cruiser, its built-in flotation compartments would not have allowed it to sink.

The cause of the mishap, apparently, came from an underwater earthquake: the wave that struck the boat, the unfortunate Burack took to be some object struck by the screw; "strong vibrations in the engine"—also his speculation; most likely, frequent jolts of the ocean floor were transmitted to the body of the boat.

Let's recall the "miracles" encountered by Columbus on his voyage to the New World: a huge wave rose up, absent any wind, and days later, light flashes appeared over the water's surface. There must be a connection between these Bermudan phenomena—in windless weather, a large wave in the open ocean can arise only from the shaking of the bottom; a marine quake, in its turn, allows for the separation of gases dissolved in bottom waters.

In the construction of a vessel, the density of ocean water is always taken into account. Any vessel is constructed with a specific store of flotation, so that the draught line (or waterline) rides above the water. Depending on construction, the degree of flotation differs among vessels. This is why in gas-permeated water Joe Tayley's vessel ended up underwater, while the powerful *Caicos Trader* merely settled closer to its waterline.

However, with the vigorous separation of a huge quantity of gases dissolved in bottom waters, if they overcome the pressure lying over the density of the water, the ocean becomes a roiling, frothing surface. And then not only the giant *Cyclops* will plunge into the ocean abyss, but even the unsinkable, outfitted with flotation compartments *Witchcraft* – just like bits of cork drown in a just-uncorked bottle of champagne, at the moment of violent separation of gases. And if the bottle has also been given a good shaking, that's "modeling" an earthquake!

Similar to water boiling in a tea kettle, the surface of the ocean acquires a white coloring. Precisely such water—white, "like milk"—was observed by Columbus, traversing the Caribbean Sea. Such water was also seen by the pilots of Flight 19, near the shores of Florida. "We're entering white water…"—were the anguished last words of Lieutenant Taylor.

X

And what about the planes?—the reader reminds me. Oh yes, the planes…The Australians' suggestion about the combustion of planes is highly dubious. The thing is, the probability of methane combusting depends on its concentration. In an enclosed space, such as a coal mine shaft, gas actually might ignite from the tiniest spark. But in the atmosphere the concentration of methane is relatively low, and drops acutely with height, which reduces to zero the likelihood of its combusting.

Besides, at the heights where planes fly there prevail strong atmospheric currents, scattering the methane into the surrounding space. If some quantity of this gas gets sucked into the plane's engine, it simply burns up along with the fuel. And completely unrealistic is the idea that a gas bubble rising from the ocean depths can "knock" a plane, flying at a height of several kilometers, out of the sky.

Part II
"If You Want to Go Crazy..."

"The first task of the mind is to distinguish the true from the false."
Albert Camus

2.1. "The Greatest Mystery in The History Of Aviation"

The reader is my witness: I invariably follow Occam's principle, attempting to explain Bermudan tragedies with natural causes. But even the most powerful underwater volcanic eruptions, just like marine quakes, cannot be the cause of mysterious aviation catastrophes, much less the traceless disappearances of planes…

From the American orbital station Skylab, some telemetric information reached Earth that operators of the space center didn't want to believe at first. The data indicated a lowering of ocean levels in the Bermuda Triangle, or more accurately in the region of the Puerto Rican Trench, the deepest in the Atlantic. But no, the radio location altimeter didn't make a mistake: over a distance of approximately 800 square kilometers, the level of the water's surface turned out to be 25 meters lower than the level of the entire ocean.

Such an unusual phenomenon, geophysicists judge, can be explained only by a gravitational anomaly. And if its dynamic force is subject to fluctuations, then it's understandable why so many different anomalous phenomena occur in the Bermuda Triangle.

…An Eastern Airlines Boeing 727 passenger jet was on its approach into Miami; radar on shore had already intercepted it, and the control tower announced its imminent landing. But at that moment, when it was almost at the very shore, the jet entered a light mist, and suddenly disappeared from the radar screens. An alarm was raised at the airport. Ten minutes later, the silhouette of the Boeing unexpectedly reappeared on the screens.

The jet landed safely. The crew members and passengers were surprised by the raised alarm; to them it seemed as though nothing out of the ordinary had happened. But, right there on the landing strip, it was discovered that the time pieces of both pilots and passengers were behind by exactly ten minutes.

This information seems so unrealistic that many commentators doubt its veracity. Berlitz, on the contrary, sees nothing impossible here.

"Although radar sometimes malfunctions," he writes, "the concurrence of time—ten minutes—is witness to the fact that the jet and passengers, for the duration of that time, were possibly somewhere in a different space-time continuum." In that incident, posits Berlitz, a role was played by powerful gravitational and magnetic disturbances, agitating near-terrestrial space upon its penetration by UFOs.

To confirm his hypothesis, Berlitz cites the results of the so-called Philadelphia experiment. It's written that the experiment was carried out in the summer of 1943 near the shoreline of Philadelphia, initiated by the US Navy and the Naval Research Office. Under conditions of utmost secrecy, the hull of the destroyer *Eldridge* was subjected to the influence of a powerful magnetic field.

The destroyer, by some mysterious process, disappeared from the dock in Philadelphia and appeared in the region of Norfolk; then it just as suddenly turned up at its own dock in Philadelphia.

I wouldn't have given these reports much credence, if I hadn't happened to discover indications of an external cause influencing the course of the planes of Flight 19.

If the route of Flight 19 were depicted by a solid line on a map, it would form a triangle, one of whose points (the beginning and end of the route) would rest at Fort Lauderdale. According to plan, by 16 hours Taylor's group should have been located near the end of the last, or third, leg of the route. Here, from a height of 1,500 meters, the Florida coast is visible even before the island of Grand Bahama, behind, has disappeared below the horizon.

But now the island has dissolved in a mist, and ahead, instead of the expected coastline there appears another group of islands. What could Taylor think? Understandably, he thought of the Florida Keys, which lie in direct proximity to the southern coast of Florida. It looked as though the planes, holding a southwest course, had somehow deviated strongly to the south.

From the testimony of Captain 3rd rank Richard Baxton, deputy commander of operations, US Coast Guard, 7th Marine district, Miami:

"…I surmise that when they thought they were over the Florida Keys, [the planes] were actually at Walter Key, within 40 miles of Grand Bahama."

And so it was: at the very time when, according to plan, the bombers were supposed to be heading in for a landing, they were still, for some unknown reason, on the second leg of their route, just north of the Bahamas.

From the testimony of Lieutenant Robert F. Cox:

"…Then I said: 'FT-28, this is FT-74. If you're over the Keys, turn your planes so the sun is on your left, and fly toward shore until you reach Miami; the next port will be Fort Lauderdale, within 20 miles of Miami. The airbase will be to your left'…"

And so, at a few minutes after 16 hours, Flight 19 took a course to the north, almost parallel to the eastern coast of Florida. It wasn't Lieutenant Cox's fault that he gave Taylor a wrong course—Cox believed that FT-28 was indeed flying over the Florida Keys.

Taylor entrusts the position of lead plane to Stivers, but as before, only open ocean lies along the path of Flight 19. And now a terrible realization begins to dawn on Taylor, that maybe the compasses on the other planes are also showing a wrong direction…

16 hours, 45 minutes

Taylor to Port Everglades:

"We will follow a course of 30 degrees (north-northeast) for 45 minutes, then turn north, to ascertain that we are not over the Gulf of Mexico."

From that moment the five bombers began to fly away from the mainland in an easterly direction.

Conclusion 17.

"The commander of Flight 19, piloting plane FT-28, was unclear as to the direction in which the Florida peninsula lay, and that uncertainty influenced his subsequent decisions."

Kusche explicates Conclusion 17 in his own way: "Taylor had been transferred to Fort Lauderdale not long before this flight, and his insufficient familiarity with the region of the Bahamas led to his erroneous assumption that he was flying over the Florida Keys…"

Lieutenant Taylor had indeed been transferred to Fort Lauderdale within two weeks of the disaster. But where did he serve previously? In Alaska? Or maybe in California? No, he had served for nearly a year in Miami. "This was a routine flight, one of many of the training flights that are performed daily at the base. All members of Flight 19 had flown this route more than once."

Taylor made decisions based on his conviction that the islands he saw were the Florida Keys. If one were to fly in a northerly direction from this group of islands, the mainland would come into view literally in minutes. But when in the course of half an hour the mainland still didn't appear, Taylor concluded that they had long ago entered the Gulf of Mexico.

And this meant: though the compasses were showing "north," some unknown power was deflecting the planes in a northwesterly direction. The fact that the pilots became disoriented was not due to their commander's unfamiliarity with the region, but to other causes, totally objective, though accompanied by unusual circumstances of the sort, by the way, of which there are quite a few in this entire tragic story.

At 16 hours, 45 minutes, the bombers bank to the east, in order to reach the Florida peninsula from its western side. Taylor still finds it hard to believe that the compasses on all five planes are giving false readings. He hands over the leadership of the squadron first to one pilot, then another, consulting about the indication of their compasses in turn. The commander is gradually seized by panic, and he changes course more frequently.

From the testimony of captain-lieutenant Donald J. Poole, Navy flight officer, Fort Lauderdale:
"When at 16 hours, 45 minutes Port Everglades established contact with FT-28, piloted by Lieutenant Taylor, I immediately asked them to tell FT-28 to take a course of 270 degrees [to the west], or simply fly in the direction of the sun. I know this message was relayed to him because I was monitoring the channel of operational communications…"
17 hours, 09 minutes

Unknown (one of the pilots of Flight 19):
"How far have we flown already? Let's turn two degrees to the east. We've gotten the hell far north instead of flying east…"
The pilots still suppose that they're flying over the Gulf of Mexico. Every minute must have seemed like an hour. The mainland didn't appear, and the planes kept pulling more sharply to the east, getting farther and farther out over the open ocean. But now the command directive (to take a course of 270

degrees) comes through: the course is reversed – Flight 19 is flying in the opposite direction. To the west!

17 hours, 15 minutes

Taylor to Port Everglades:

"I copy you, thanks very much. We're now flying on a course of 270 degrees."

17 hours, 16 minutes

Taylor:

"We'll fly on a course of 270 degrees until we reach shore or until we run out of fuel."

17 hours, 22 minutes

Taylor to the pilots of Flight 19:

"When someone is down to 10 gallons of fuel, we all land on the water together. Everyone got it?"

The command directive—to hold a course of 270 degrees—told Taylor that the squadron was actually to the east of the Florida peninsula. At 17 hours, 50 minutes, a fix was finally made on the location of Flight 19, exactly in the "quadrant" where two hours later the Martin Mariner search plane disappeared. Flying out of Banana River, the hydroplane reached that location in 23 minutes. If Flight 19 had continued to fly in its set direction, then in 15 – 20 minutes the panorama of the lit-up Florida coast would have opened up before them.

Why didn't the planes reach shore? Did they change their direction again? But Taylor's decision to follow a course of 270 degrees "to the end" puts that proposition in doubt. To think that the planes made a water landing around 18 hours also doesn't seem right—they had enough fuel for another two hours of flight. Only one possibility remains: though the compasses showed "west," some unknown force was deflecting the planes from their specified course, inexorably, fatefully drawing events to their tragic conclusion.

2.2. Was Lieutenant Taylor Drunk?

In a like manner, regarding Taylor's reluctance there later emerged all sorts of conjectures: the lieutenant didn't explain why he didn't want to participate in the training flight, and to many this seemed strange. (Could he have had a premonition?) Among others, there was the speculation that Taylor possibly didn't feel well, or wasn't sober.

If the leader of Flight 19 was experiencing some sort of ailment, he had no reason to hide it. Did Taylor show up for work that day in a drunken state? Definitely—no. As indicated by Curtis' testimony, there wasn't anything strange about Taylor's behavior, he conducted himself normally in all respects.

And now a question for you, esteemed reader: doesn't it surprise you that Lieutenant Taylor, literally an hour before takeoff, asked to be replaced by a different instructor pilot? Are you having trouble coming up with an answer? Let's turn to another fact.

From the report of the investigative commission:
Facts 8 – 12.
"In each plane except Gerber's there was a crew of three, including the pilot. In Gerber's plane the crew was one person short."

Why was the crew one man short on Gerber's plane? Perhaps that man was sick? But it's possible to suggest something else, for example that this crew member, pleading illness, in reality possessed information about preparations of sabotage. However, why guess—the security service probably had a talk with him. It turns out that this member of Gerber's crew, Allen Kosnar by name, "requested to be relieved of the flight for reasons of intuition." And his request was honored.

Excuse me, but why is Lawrence Kusche silent about this fact in his book? Apparently from a desire to conceal from the reader another riddle: a premonition of disaster arose in two people simultaneously. And the disaster took place. Could it be clairvoyance? Or telepathy…

One inadvertently recalls stories about the Philadelphia experiment when, under conditions of influence from a strong magnetic field, a heightened PSI factor was observed in the sailors on the experimental destroyer.

A similar field, probably, awakened in Kosnar and Taylor a "dormant" faculty of foresight, a premonition of mortal danger. Judging by everything, uncommonly strong gravitational and magnetic disturbances, arising in the region of the Bermuda Triangle, also encompassed the Florida peninsula a few hours before the takeoff of Flight 19.

According to contemporary cosmological theory, there exists a whole array of parallel, isolated from each other, universes. They all, like our own universe, are filled with stars and planets, on some of which one may suppose there also exists rational life. These neighboring universes are literally penetrated with power fields, and apparently, it's by these that they are fenced off, or "locked."

However, with the warping of space, it's as though windows are opened up through which these highly developed residents of parallel worlds enter or leave our universe in their "flying saucers."

There are publications in which an idea is established concerning a possible dynamic connection between psychic phenomena and structural peculiarities of the time-space continuum. It's suggested that the warping of nearby space activates hitherto latent parapsychological abilities in humans: in the generated wrinkles of physical space they begin to see, as in a mirror, the reflection of future events…

It's possible that psychic disturbances emerged in Taylor later, during the fateful flight. The ground for this supposition was provided by the report of a journalist who devoted several years to investigating the riddles of Flight 19. A "stunning" announcement was made by him on American television in 1975. It seemed that, as the drama was unfolding, a ham radio operator heard Taylor's words: "Don't follow me…they look like aliens from space…"

The announcement was investigated. Yes, agreed the foes of mysteries, the words "don't follow me" were definitely spoken by the commander of Flight 19, but as for "aliens from space," this is a fabrication of sensation mongers.

There were arguments over the first part of the phrase as well. Thus, Berlitz is convinced that Taylor's words—"don't follow me"—are addressed to the pilots of Flight 19. No, object the opponents, that was Taylor's answer to

Lieutenant Cox, who wanted to fly out to meet the lost planes and conduct them to their base.

However, this explanation is easy to rebut. Lieutenant Taylor was responsible for the lives of the 13 members of the flight group he led. It's impossible, in a situation when the pilots had lost their orientation, that their commander would have refused proffered help. It's clear that Taylor's words—"don't follow me"—were addressed to the pilots of Flight 19…

X

Joan Powers of Mount Vernon (New York) woke abruptly in the middle of the night. A nebulous anxiety caused the woman to do something she'd never done before—telephone the airbase where her husband, John Powers, was stationed. The jury officer haltingly answered: "We are unable to call your husband, Captain Powers, he's on a flight…"

That her husband, together with the other pilots of Flight 19, did not return from their mission, Joan Powers found out only the next morning through an expedited news release.

2.3. Those Inclined to Believe in UFOs

In April of 1988, the Canadian journal *Weekly World News* proclaimed sensational news: an airplane was discovered on the moon. "We have photos," the journal claimed, "which show an American bomber of World War II vintage, lying in a lunar crater." Later, it's true, there appeared an announcement about how the photograph of the plane in the lunar crater was an artificial photo montage. But has there ever been any evidence of teleportation or extraterrestrials, that wouldn't have been met with skepticism?

On January 30, 1948, the passenger airliner *Star Tiger* disappeared in the Bermuda Triangle. The circumstances surrounding this event were so mysterious that an investigative commission noted in its report: "Never has an investigation been faced with solving a more difficult riddle." Because the airliner's catastrophe couldn't be explained by technical malfunctions, the commission suggested that the cause of the tragedy must have been external.

Kusche: "Those inclined to believe in UFOs lend particular significance to one phrase in the conclusion of the commission's report, where it says that 'it's possible for some sort of external cause to operate on both man and machine'."

Obviously, they consider that the commission assigns responsibility for the demise of the aircraft to UFOs, but lacks the courage to make the pronouncement. However, within the context of the report, the words "external cause' may refer to anything at all; for example, we might speak of unfavorable weather conditions..."

From the report of the investigative commission:
"In the region where the *Star Tiger* was located at that moment, as well as along the route of its impending flight, the weather continued calm, without thunderstorms or deluges of any strength, either upward or downward, that could destroy the airliner."

I would give a lot to know what external causes, besides weather conditions, Kusche could have named.

Miami *Herald*
Monday, June 7, 1965
"On Saturday a massive search was initiated for a military plane known as the 'Flying Boxcar' with a crew of 10…The Coast Guard surmises that the C-119 disappeared in a region south of the Bermuda Islands, *approximately 280 miles from Miami.*

"This gigantic twin-engine plane flew out from the Homestead airbase on Saturday at 19 hours, 47 minutes…A spokesman for the Homestead airbase stated: "…since yesterday evening, we've had no word of the aircraft." The C-119 was due to arrive at Grand Turk at 23 hours, 23 minutes on Saturday. At dawn a search began, covering a region of over 2,000 square miles of the Atlantic."

Miami *Herald*
Tuesday, June 8, 1965
"The lost US Air Force plane was only 45 minutes away from arriving at its destination, when it mysteriously disappeared…Flying over Crooked Island in the southern portion of the Bahamian archipelago, it made its last contact at 23 hours, 00 minutes. At that time the heavy twin-engine C-119 was some 100 miles from the airstrip on Grand Turk. As stated by a US Coast Guard representative, 'There was nothing in the message from the C-119 that could have caused any anxiety, but we didn't hear from it again…'"

By Monday the search had expanded to cover a region of up to 100,000 square miles; this is the region known to veteran pilots as the Bermuda Triangle…Precisely here, during the years of World War II, hundreds of planes, vessels, and submarines went to the bottom…Later on, a quantity of planes also vanished under mysterious circumstances.

"'It's strange,' commented an experienced pilot out of Homestead, who had flown here during both war time and peace, 'that when planes perish in the region of the southern Bahamas, there's no trace of them left.' The same transpired with the vanished C-119, not a single fragment, no signs of life…"

As usual, Kusche prefaces his description of the event with the Legend. At the same time, he's forced to admit that in the given instance the Legend concurs with newspaper reportage about this incident. And the newspapers published (in 1973) material from the international Center for UFO Studies in which is stated, in part: "It would not be surprising if the C-119 had been snatched by a UFO."

"In the very same days when the C-119 disappeared, the American spacecraft Gemini 4 was orbiting the Earth, and astronaut James McDivitt noticed a UFO with something resembling arms. A few minutes later, McDivitt and another astronaut, Ed White, observed a second UFO over the Caribbean Sea…An examination of the images taken by the astronauts showed that the unknown flying object noted by them had nothing to do with the spacecraft…"

On June 4 McDivitt really did see some sort of object with long "arms" stretched out in all directions. For several days, programmers never left their computers, trying to ascertain what kind of object had appeared approximately within 10 miles of Gemini 4.

At first, suspicion fell on Pegasus 2, a gigantic satellite with 96-foot, arm-like antennas; however, subsequent calculations showed that Pegasus was at that time over 1,000 miles from the given point. McDivitt described the object he'd seen, indicating that it was white and cylindrical, with an "arm" flung out sideways.

Kusche writes a letter to McDivitt in hopes of getting from him some augmenting evidence that could help dissipate speculations about machinations of extraterrestrials. The answer, however, turned out not to be in the spirit of the foe of mysteries: "In answer to your letter of January 22, I wish to inform you that during my flight in Gemini 4 I really did see what some people call a UFO…

"The object that I saw remains unknown. But from this it doesn't follow at all that it was a spaceship from some distant planet somewhere in the Universe. It also doesn't follow from this that it wasn't a spaceship. From this only one thing follows: during my flight I saw an object that neither I nor any other person has yet been able to identify or determine, what it is."

X

On December 28, 1948, a DC-3 passenger airliner was headed out of San Juan (Puerto Rico) to Miami. Captain Robert Linquist and his co-pilot, Ernest Hill, were responsible for 27 passengers returning home after the Christmas holidays. Not long before departure it was discovered that the radio transmitter wasn't working, due to discharged batteries.

"Learning that a recharge of the batteries would take several hours, Linquist directed the mechanic to add water to the batteries and take them back to the plane, without spending time on recharging them…It was agreed that, until the aircraft's generators gave them enough energy for satisfactory performance from the transmitter, the aircraft would remain in the vicinity of San Juan."

Eleven minutes after takeoff, Linquist called the control center at the airport in Miami. The control center didn't hear the call for some reason, but it was intercepted by the communication center of the bureau of civil aviation in San Juan. Interference with radio connection didn't stop there.

In spite of numerous attempts, the communication center wasn't able to contact the DC-3 again. During the course of the flight, numerous radio stations intercepted messages, but direct contact with the aircraft for updating its location along the route was never established.

At 04 hours, 13 minutes Linquist called Miami, said all was well on board, and requested instructions for a landing approach. As in the first instance, the radio transmission was not heard by its addressee, but was intercepted in New Orleans. This was the last message from the DC-3.

An urgent search was begun a few hours later. Hundreds of vessels and planes combed the ocean along the entire length of the route from San Juan to Florida; the scope of their search included the Caribbean Sea, the Gulf of Mexico, the Florida Keys, Cuba, Haiti, and the Bahamas. At the place of the supposed demise of the airliner, no trace of catastrophe was found.

From the report of the investigative commission:

"It's known that the radio transmitter on the plane was working when New Orleans intercepted the message about its position…It's possible that the pilot incorrectly determined his location…Since there was enough fuel for 7 ½ hours of flight, and the last message was received after 6 hours, 10 minutes of flight, a miscalculation of position could have proved fatal."

Kusche: "Although the weather was fair, it nevertheless could have been one of the factors leading to catastrophe. Weather forecasters in San Juan informed Linquist that at the start of his flight there would be a weak southwest wind, which would then change direction and blow form the northwest.

"Adjusting for wind, Linquist had to steer his plane a bit to the left of his specified course. However, as they got closer to Miami, the wind changed direction again and blew form the northeast. If the pilot didn't know this, then, even though the wind wasn't strong, it could have deflected him off course…"

All by itself, the wind (especially a weak one) cannot cause a significant deviation from course. This means that, if we're talking about a considerable deviation from course without the pilots being aware of it, it may be supposed that the readings of onboard navigational instruments don't correspond to the actual course of the aircraft.

Kusche: "During the 1940s, when radio apparatus essentially worked on low frequencies, determining one's position over the ocean was far from simple…Until the appearance of omni-directional beacons and special television equipment for measuring distances, pilots didn't have sufficiently reliable methods of gauging their location in the air…When Linquist said that he was within 50 miles south of Miami, he was merely giving approximate information about the location of his aircraft…"

These lines must be understood thus: if the DC-3 (just like the *Star Tiger* and the planes of Flight 19) had been equipped with navigational apparatus not of the 1940s but, say, the 1980s, the disaster would not have occurred. I'm going to allow myself to express some doubt about this and posit that the cause of the tragedy was not technical, but "external."

Such a proposition, however, may confirm, and even lift to the rank of hypothesis, a similar incident (an unintentional deviation from course) with an aircraft of the 1980s, whose navigational equipment functioned on high frequencies. In the words of the Hungarian mathematician George Pólya, "a proposition becomes more plausible when an analogous proposition becomes more plausible." This logical procedure in known as a method of proving the instrumentality of the analogy.

2.4. The Mystery of Flight 007

On August 31, 1983, a Korean Airlines Boeing 747 passenger jet, following a route from New York to Anchorage to Seoul (flight 007), strayed off course and intruded into Soviet airspace. Here it was destroyed by the Soviet air defense. 269 people perished.

The reason for the deviation off course may have been revealed by recordings contained in the "black box" (flight recorder)—in it were secured all the flight data, including conversations of the pilots between each other and the ground. But the "black box" sank in the ocean, along with the plane, somewhere near the southern shore of Sakhalin Island. Western observers then expressed the conjecture that it was found by Soviet divers, but Moscow denied this.

A few days after the tragedy, a journalist asked the USSR Minister of International Affairs, A. A. Gromyko, "Do you think that the world will soon forget the story about the Korean airliner?" Gromyko calmly replied, "I'm sure of it." And so it probably would have been, if over the ensuing years, one after another, there hadn't come to light circumstances making the incident more and more mysterious. No wonder it's sometimes still discussed in the media and on television programs in various countries.

In November of 1992, during his visit to South Korea, Russian President Boris Yeltsin handed over to the government of that country the "black box" from the Boeing 747 passenger jet. There was hope that now some facts explaining the deviation off course would be made public. However, not two weeks later Seoul informed Moscow that the flight recorder contained almost no information. Either part of the magnetic tape had been deliberately erased during the Gorbachev administration, or these recordings were damaged by prolonged submersion underwater.

The half-erased tapes—these are an obvious riddle. But other mysterious (peculiar!) circumstances may be traced throughout this incident, to which experts didn't pay sufficient attention at the time. Taking these circumstances into account, the versions explaining the reason for the jet's deviation off course turn out to be highly vulnerable…

According to international aviation law, an aircraft intruding on the airspace of another sovereign government is subject to interception. Interception, in its turn, implies forcing the intruding aircraft to land, using universally accepted procedures. The country that destroys an intruding aircraft is obligated to provide evidence that all internationally stipulated measures aimed at interception had been exhausted. No reasonable explanations were forthcoming in this case. And there ensued a worldwide scandal.

Rome: "Defense Minister Spadolini stated that the attack on the passenger Boeing was an irrational act, filling civilized people with horror and constituting a dishonor on its perpetrators."

Brussels: "The Minister of International Affairs severely condemned the cynicism of the Soviet attack on unarmed people."

Seoul: "The President of South Korea accused the USSR of the most barbaric crime possible."

Bonn: "The West German government condemns the actions of the Soviet authorities, deeming them as absolutely unacceptable in human society for their brutality…"

In all fairness, it must be noted that the leadership in Moscow was not responsible for the tragedy itself. The decision to destroy the Korean airliner was made not only without the knowledge of the Politburo, but even without the authorization of then-USSR Minister of Defense, Marshall Ustinov.

This is why news about the shooting down of a passenger airliner shocked the Moscow leadership, which began to categorically deny the incident. But already on the second day after the tragedy, the Armenian newspaper *Red Star* admitted that the jet had been shot down by mistake.[13]

Then the head of the general staff of the Soviet Army, Marshall N. Ogarkov, stated: "No mistake was made." A few days later the head of the senior staff of Air Force troops, General-Colonel S. Romanov, attempted a justification:

---

[13] In the article "Pilot admits error…" the blame for the tragedy is placed on the pilot of the Soviet fighter jet that shot down the airliner. This confession appeared so out of the ordinary to Soviet readers that the newspaper edition sold out in a matter of hours.

the airliner didn't react to signals from the Soviet fighter jets, so they had to attack it.

Observers in the West noted with surprise that such a lack of coordination in the statements of Soviet representatives and press organs hadn't been seen for a long time. If, as stated by Marshall Ogarkov, there had been no mistakes, if Air Force units acted according to instructions, why then were several high-ranking officers of the Far East headquarters removed form duty?

Why was the commander of the Soviet Air Force, Marshall Koldunov, contrary to established Kremlin tradition, denied the usual medal commemorating his sixtieth anniversary? Why was General-Colonel Romanov, along with his deputy, Chesnokov, demoted in rank?[14]

All of this indicates that Moscow covertly condemned the local command. Apparently, the Air Force officers in the Far East violated the rules of the System—ventured to take responsibility at the highest level without consulting the political leadership of the country. But why? They couldn't possibly not understand how fraught the consequences of destroying a foreign passenger airliner would be!

If we analyze the circumstances preceding the shooting down of the Korean Boeing 747, and compare them with the circumstances preceding the disappearance of Flight 19, it's possible to discover that some of them are surprisingly similar.

Thus, we have at our disposal a unique opportunity to verify the proposition about external factors deflecting a plane from its specified course. But first we'll attempt to make sense of the political and diplomatic nuances that tie the history of the Korean Boeing passenger jet into a tight and extremely tangled knot.

---

[14] Romanov was transferred to the German Democratic Republic, where he soon perished, as was announced, "in the course of fulfilling his service duties."

2.5. Devilry in the Soviet Skies

The KAL Boeing 747 passenger airliner lifted off from New York on August 31 at 0 hours, 25 minutes, and touched down at the airport in Anchorage, Alaska.[15] An hour later it continued on its journey. The route of Flight 007 (Anchorage – Seoul) passes close to the eastern borders of the USSR; at the Kuril Islands they are separated by only 50 kilometers.

Flights departing from Anchorage customarily adjust their course by the airport's radio direction-finder. It happened that, right at that time, the VHF Omnidirectional Radio Range (VOR) beacon at the Anchorage airport was out of order, with repairs requiring several days.

A defect in one of the onboard course pointers was also discovered. But these were just trifles: the captain, Chan Ben In, was an experienced pilot; over his 11 years of service at KAL he had flown this course many times. Additionally, there is another VOR beacon at Bethel, 500 kilometers out of Anchorage. Here pilots verify the course set in Anchorage.

"Ten minutes after takeoff, KAL-007 begins to slightly deviate off course to the right. After the first 300 kilometers of flight, the deviation consists of nearly 11 kilometers...The observer on duty at the Federal Aviation Administration (FAA) radar station doesn't consider it necessary to inform the crew of this fact."[16]

Maybe the observer on duty figured that the crew, once they entered the radar zone at Bethel, would correct their course anyway? Could be. *"But for some reason neither Chen, usually so methodical, nor his co-pilot, Son, carried*

---

[15] Time indicated here and below is local time.

[16] Text in cursive represents citations from an editorial in the magazine *Country and World*, No. 1 – 2, 1985, Munich. The article reflects the opinion of official Washington

this out. Son, making contact at 49 minutes out of Anchorage, reports that everything is fine, and the plane is now directly over Bethel."

"The situation is now serious. If the pilot, known to be extremely accurate (to within one meter!) with his beam transmitter, believes he is directly over the specified object when he is actually 25 kilometers off course, it must mean that he has lost his orientation capability."

Lost his orientation capability or intentionally strayed off course? –this is the point at which contrary opinions have crossed paths. It's from this point that two notorious versions take their origin—Soviet and American. We'll turn to these versions later, but now let's see how events played out.

05 hours, 31 minutes
By this time, KAL-007 is 185 kilometers off course.

06 hours, 57 minutes
The deviation stands at 450 kilometers. As before, the crew doesn't notice anything.

08 hours, 07 minutes
The airliner intrudes into Soviet airspace, flying over Kamchatka.

American radar, mostly situated along the Soviet border, for some reason doesn't notice the intruder. This "for some reason" is **the first special circumstance**. Sent to intercept, "Soviet fighter jets took off from the airport at Petropavlovsk too late." This is probably why the local command didn't inform Moscow of the incident.

After all, Moscow leadership will likely want to know the reason for the tardy observation of the intruder. This must mean glitches in military procedure! This must mean epaulets will fly off of somebody's shoulders! No, better to first intercept the intruder, then inform Moscow. Victors aren't judged!

*"However, the unforeseen occurs. Either from lack of experience, or from the high, thick cloud cover, the fighter jets are unable either to catch up to, or even see the giant airliner…***(the second special circumstance)**. *After all, this*

isn't just a "snafu," an accidental error, a setback; this is evidence. Evidence that the air defense of the USSR is unreliable…"

"Everything went haywire that morning. Ground services sent the fighter jets where KAL-007 was not, and couldn't possibly be…"—**the third special circumstance**. *As the giant airliner proceeded serenely ahead…general chaos and panic increased. Recorded discussion between pilots, ground services, and staff is witness to this."[17]*

Just don't panic! Radar indicates the location of the intruder, but the group of fighter jets can't seem to find it. Some kind of mystery…Meanwhile, KAL-007 crossed Kamchatka and, unimpeded, entered international airspace over the Sea of Okhotsk.

"From recordings of conversations between Soviet Air Force officers, it follows that the moment has passed. It's clear that the 'spy,' who managed to fly over a secret base on Kamchatka, won't tempt fate twice, and will calmly move along the Kuril Islands toward Japan. Undoubtedly, there's logic in their calculations. But KAL-007 operates by the laws of a different logic, the logic of error."

The Korean airliner once again nears the Soviet border, this time at the southern tip of Sakhalin Island. At this point, local staff should have realized that something aboard the plane wasn't right; doubts should have arisen: is this really a spy? But there's no time for second thoughts. In the Sakhalin region an air alarm is declared. From several airbases, one after another, interceptor fighter jets take off.

"Soviet pilots and ground command fear most of all letting KAL-007 go. This is why the fighter jets, without waiting for the airliner to appear over Sakhalin, fly to meet it in international airspace…The situation on Sakhalin is

---

[17] Heightened activity on the Soviet border attracted the attention of personnel on American and Japanese bases, situated on the opposite side of the strait. From that moment, the incident was monitored by powerful American and Japanese installations. Later, the American government released the audio recordings of discussions between services of the Soviet Air Force.

strained to the limits. It's clear to Air Force officers: if the intruder now escapes unpunished, their careers are finished."

Now Soviet interceptors seek the intruder in neutral skies. And with only 4 minutes of flight remaining until the state border is reached, Su-15 interceptor pilot, Major Gennady Osipovich, declares to his command station: "I have a visual, and see it on my screen."

"Two hours had already passed since the object was first picked up on radar, and only now was visual contact made." **(The fourth special circumstance.)** The head of the general staff, N. Ogarkov, explained later that this was due to the interception being conducted in conditions of poor visibility. However, this was a facetious explanation, since all Soviet fighter jets are equipped with radar.

"A minute after confirmation: 'I see him…' Three minutes separate KAL-007 from Soviet airspace. And after 20 seconds: 'Target doesn't reply to query.'"

The fact that not one of the queries make by the Soviet jet fighter was heard aboard KAL-007 represents **the fifth special circumstance**.

"…In another 14 seconds Osipovich reports to his command station that he has engaged his rockets…In those very seconds, the airliner has crossed the invisible border and is once again in the airspace of the Soviet Union."

To ground command, and to Major Osipovich, it's now clear: the airliner is not a spy, its pilots don't even suspect that they're off course. But what to do, if literally in a few minutes the "intruder" will once again leave Soviet airspace. Now it's too late to ask Moscow how to further proceed.

"…With four minutes until the international border, Osipovich hesitates for two of them…And only in the final seconds, as the airliner approaches the shore, does the pilot make up his mind. 'I'll now attempt with the rockets!' At 03 hours, 26 minutes, 22 seconds, Osipovich declares: 'The target is eliminated.'"

2.6. "If You Want to Go Crazy…"

One warm summer evening in 1970, two likable young men wearing the uniform of the US Air Force were strolling in a park in the vicinity of Colorado Springs. Suddenly they were approached by an unkempt-looking man, who could have been taken for a panhandler.

"You know," said the stranger, "during the war I was also an officer. Except I served in the Navy. They pulled me into an escapade there, and then they got rid of me. They said I'd lost my mind. But I simply couldn't stand the strain. So they threw me out." And in confirmation of his words the stranger drew from his pocket and showed the young men a worn identification card with the seal of the US Navy.

The officers thought the stranger was about to ask them for money. And if he got it, he would continue his story. But this didn't happen. "They wanted to make a ship invisible," the stranger continued, after a pause. "Can you imagine what a splendid camouflage, if everything succeeded? Actually, it did succeed. With the ship, I mean. But we, the crew…Something didn't work with us. We simply couldn't withstand the effect of the force field…"

"But what are you talking about?" interrupted James Davis (one of the young men). "Was this some sort of experiment?"

"Electronic camouflage," nodded the stranger. "Some sort of electronic camouflage, achieved with the aid of pulsating force fields. I sure don't know what kind of energy they were using, but the strength was brutal. And we couldn't bear it, not a single one of us. Though the effects were different for everybody. Some had double vision, others were laughing and staggering like drunks, and someone else fainted.

"Just imagine, some of them insisted that they found themselves in another world, where they saw strange extraterrestrial beings and communicated with them. Some died. But I was one of the survivors…They just wrote us off. Like mentally unbalanced and unfit for military service…First, of course, they isolated us for a few months—'for rest and relaxation,' as they put it. And also, supposedly, to get it into our heads that nothing of the sort ever happened to us. In any case, in the end we were all sworn to secrecy. Though after all, even without that, not a single person would have believed such a story. But you believe me, don't you?"

The young officers, of course, didn't believe him. They kept waiting for the stranger to ask them for money, but he didn't ask, just quietly moved on. Most likely, he was a mentally ill person...

In 1975, the first book about the Bermuda Triangle by Charles Berlitz was published. The book ended up in the hands of the abovementioned James Davis. Imagine his astonishment when he found in it a description of the Philadelphia experiment! After some thought, Davis contacted Berlitz by telephone and told him about the strange person and the story he told.

Berlitz was very grateful to Davis for the information—the disappearance of the ship in front of a crowd of people was evidently not a fabrication. And to the extent that analogous physical effects (magnetic disturbances, the disappearance of people and technology) are often noted in places where UFOs also appear, then it's highly likely that events in the Bermuda Triangle are caused by the activity of an extraterrestrial mind.[18]

It's too bad, of course, that Davis didn't think at the time to question the stranger more closely, or get his name and address. Now he was unlikely to be found. But the name of another possible witness of the Philadelphia experiment was known – Carlos Miguel Allende...

In August of 1955, a parcel addressed to Admiral N. Fert, head of the Office of Naval Research, arrived without a sender's name or return address.[19] When Major Ritter opened the package, he found a recently published book, *The Case for the UFO,* by the astrophysicist Morris Ketchum Jessup. There was no accompanying letter, and the Major's first impulse was to toss the book into a waste basket.

But suddenly his attention was drawn to the numerous comments in the margins, made with colored pencils. These comments addressed the

---

[18] In the book by Charles Berlitz titled *The Roswell Incident,* sightings of UFOs are summarized, and the idea of their extraterrestrial origin is expressed.

[19] The Office of Naval Research (ONR) is an organization within the United States Department of the Navy responsible for the science and technology programs of the US Navy and Marine Corps.

mysterious disappearances of ships, planes, and people—mostly in the region of the Bermuda Triangle. They also touched on unusual storms and cloud formations.

The comments in the margins were so interesting that the author of the book was invited to Washington. As an employee of the Office of Naval Research put it, "The author of the comments had extensive information about 'creatures from UFOs,' about extraterrestrial phenomena, and many other things discussed only by psychiatrists and people involved in cults and mysticism. And the main thing wasn't whether they actually exist. More important was the author's amazing familiarity with these issues."

But these weren't the only things puzzling Morris Jessup. A few months earlier, he'd received a letter of the utmost strangeness. The letter claimed that American secret services had deliberately led world society astray regarding the last theoretical investigations of Albert Einstein.

Truly, the work of formulating the theory of a unified field, by which the great physicist wished to express the interconnectedness of the forces of nature, had been completed. The results of the Philadelphia experiment supposedly attest to that. Einstein, according to the author of the letter, had simply destroyed his notes, guided by humanitarian considerations. This could be confirmed by Einstein's close friend, the English philosopher Bertrand Russell, who felt that humanity had not yet evolved far enough for such an experiment.

The letter was written with colored pencils. It was signed—Carols Miguel Allende. Jessup then wrote to the author of the letter, asking him for details. But he never got a reply. Now Jessup expressed the conjecture that the comments in the margins of his book were apparently made by the same Carlos Allende.

Morris Jessup made a few more trips to the Office of Naval Research in Washington. The reasons for this were the following three letters from the mysterious Allende. They were primarily regarding the results of the Philadelphia experiment. I'll cite some excerpts from these letters, containing references to phenomena similar to Bermudan phenomena.

"...The result was complete invisibility of the destroyer vessel and its entire crew (October, 1943). The magnetic field was in the form of a rotating ellipsoid extending for 200 meters...on either side of the ship. Everyone who was in that field had only a blurry outline, but could perceive all those who were aboard the ship...like they were walking or standing in the air. Those who were outside the magnetic field couldn't see anything at all, except the ordinary trace of a ship's hull in the water…

"If you want to go crazy, access this information. Half the officers of that ship's command are totally insane. Some of them are still held in suitable institutions where they get qualified professional help when they 'levitate,' as they themselves call it, 'levitate' or 'get stuck.' In that state they're unable to move of their own free will, unless one or two companions that are in the magnetic field with them don't quickly come over and touch them, because otherwise they'll 'freeze.' Once a person 'freezes,' his location is thoroughly camouflaged…

"The 'frozen' one usually loses his mind, raves, and babbles nonsense, if the 'freeze' has lasted more than one day by our way of counting time. I speak of time, but…the 'frozen' ones perceive time differently than we do. They resemble people in a twilight state, who live, breathe, hear, and feel, yet aren't aware of much, as though they exist, only in the next world. These don't perceive time like you or I do…

"There aren't many crew members left who took part in that experiment…Most of them went insane, one simply disappeared 'through' a wall in his own apartment, in front of his wife and child. The experiment, as such, was an absolute success. But it had a weird effect on the crew…Check out the observation ship, the *Andrew Furuseth* (Matson Co., port of registration, Norfolk. Perhaps the company has a watch journal of that voyage, or perhaps the Coast Guard has it.)

"…As a result of cold and sober analysis, I want to inform you and the field represented by you of the following.

"The Navy didn't know that people could become invisible if they were…under the influence of the field.

"The Navy didn't know that people could die from the side effects of the field…

"What's worse, and is never mentioned: when one or two people, visible to everyone within the field, just went off into nothing, and nothing tangible was left of them (neither when the field was activated, nor when it was deactivated), when they simply vanished, the fears developed. Things got worse when one who seemed to be visible departed 'through' a wall of his own house. The vicinity was thoroughly inspected with the aid of a portable field generator, but not a trace of him was found. Then the fear increased to a level where none of these people, or the people who were conducting the experiments, could continue them.

"I also want to mention that the experimental ship vanished from its own dock in Philadelphia, and a few minutes later appeared at different docks in Norfolk, Newport News, Portsmouth. It was clearly and precisely identified, but then disappeared once again, and a few moments later returned to its own dock in Philadelphia. This was also in the newspapers, but I don't remember where I read it, when it happened. Possibly during the time of the later experiments. Possibly also in 1946, after the experiments were halted…

"I think, if you had been working then with the group that took part in the project, and if you had known then what you know now…not one of these incidents would have taken place. Really, they could have been prevented, in part, by implementing a more forgiving program and a more scrupulous selection of officer cadres and crew. But this didn't happen.

"The Navy simply used the personnel who happened to be available, barely considering the character and individuality of that materiel. With care, great care, in the selection of the ship, officers, and crew, with enough attention to such ornaments as rings and watches, as well as identification tags and belt buckles, I think to some extent it would have been possible to successfully dissipate the ignorance, resulting from fear, that surrounds this project.

"Personnel management documents of the Navy at Norfolk will show who was assigned to the ship *Andrew Furuseth* at the end of September or in October of 1943…I leave it to you to decide whether this is worth a large amount of work or not, and write in the hope that it will be accomplished."

"I hope you understand that their failure resulted not in the implementation of metallic and organic invisibility, but in the implementation of involuntary transport, in the blink of an eye, of thousands of tons of metal, along with people.

"Unintentionally, and to the great consternation of the Navy, this has already happened once with an entire ship and its crew. I read about this, and also about the actions of sailors who left their bases without leave, and who at that point were invisible, in one of the Philadelphia dailies…these people have a very high PSI factor, with could intensify…"

Although the content of the excerpted letters is in many ways aligned with the ideas laid out in the book *The Case for the UFO*, Jessup nevertheless (or maybe precisely because of it) couldn't shake the thought that someone was messing with him in a most sophisticated way. Still, the Office of Naval Research in Washington took a serious interest in these letters. And that's significant!

2.7. In the Role of Sherlock Holmes

The name of Carols Miguel Allende long remained surrounded by mystery, and as time went on it became legendary. This person knew too much. He was hunted by an army of journalists, but he kept changing addresses, and it proved impossible to find him. Then the inquisitive investigators took the path of least resistance, deciding to verify how closely the testimony in Allende's letters regarding the Philadelphia experiment squared with reality.

They sent queries to the firm Varo Corporation, a wartime manufacturer of powerful magnetic force field generators. They received an unusually terse reply: "The firm is not interested in discussing this subject with you or anyone else…All your queries, letters, and telephone calls will go unanswered."

The Office of Naval Research answered: "The ONR did not, in 1943 or at any other time, conduct research regarding invisibility."

One of two possibilities: the US Navy classified the information regarding the Philadelphia experiment, or such an experiment never took place. Moreover, Allende didn't seem at all like a con artist, since, unlike people seeking personal profit, he didn't solicit any meetings with those to whom he addressed his letters.

Was he insane? Doubtful, since the content of his letters testified to the consistency of their author's thinking, showing him to be highly informed about the substance of his topic, and possessing a breadth of knowledge as well. That Allende had selected Morris Jessup and the ONR as his addressees held a logic of its own: the physical effects characteristic of UFOs are in many ways analogous to those manifested in the Philadelphia experiment.

It was also possible to understand from a humanitarian perspective the motives that moved Allende to make the corresponding warning "This dangerous experiment may be attempted by ignorant and evil people!"

And yet, strange circumstances came to light in the attempt to obtain information about the ship *Andrew Furuseth*, whose crew members, if Allende were to be believed, witnessed the disappearance of the destroyer DE-137 (the *Eldridge*). The watch journals from this ship "disappeared," while provided information regarding its location in the summer of 1943 proved erroneous. ("Mistakes" were discovered 30 years later, after the Navy declassified certain material.)

But in documents from the shipping company Matson, some extremely valuable information came to light: on August 13, 1943, the *Andrew Furuseth* took on some sort of cargo in the port of Norfolk; the list of deck crew members featured the name of Carlos M. Allende.

The destroyer *Eldridge* entered service on August 27, 1943. Both ships were part of a wartime mission off the north shores of Africa. And if the expedition was connected with the camouflage experiment, Carlos Allende could very well have been a witness. It's true that at first the investigators had their doubts as to why the Navy would decide to conduct a secret experiment in full view of the crews of other ships.

Later, though, this situation was explained by testimony received from an "authority," who had been engaged in the Navy's radio location program.

This is what the "authority" said: "I think they carried out several trials…with the goal of determining the influence of a strong magnetic field on location installations…My conjecture (I repeat—conjecture) is that the entire receiving apparatus was placed on different ships…to determine what happens 'on the other side' when passing through a field of radio or radar waves. Undoubtedly, scientists had to be monitoring what influence the field would have on visible light…"

One more piece of information came from Mr. Patrick Macy, who served in the Navy as a documentary filmmaker during the Second World War. "In Washington, in 1945," recalls Macy, "I chanced to see part of a film, about an experiment conducted at sea, that was being shown to the top Navy brass. I only remember parts of the film, since I was on duty and couldn't just sit and watch like the others…But I remember it was about three ships. They showed how two of the ships were pumping some sort of energy into the third one, which stood between them."

"At the time I thought they were radio waves, but I can't say anything definite, since obviously I wasn't in the loop about any of this. After some time, the third ship—a destroyer—began to gradually disappear in a kind of transparent fog, till all that was left was its trace in the water. Then, when the field or whatever it was, was switched off, the ship reappeared once again out of a thin mist. This was apparently the end of the film, and I happened to overhear some of them discussing what they saw. Some were saying that the field was turned on too long, and that explains the problems that developed with some of the crew members."

"An incident was mentioned, for example, where supposedly someone from the crew simply vanished while having a drink at the bar. Someone else said that the sailors are still not in their right minds. There was also talk about how some sailors disappeared for good. The rest of the discussion went on too far away for me to hear it…"

Documents of the General Services Administration in St. Louis testify: during 1943 – 1944, Albert Einstein worked at the Maritime Administration in Washington, D. C. as a scientific consultant. And if the Philadelphia experiment isn't a fiction, it's unlikely that it would have taken place without Einstein's participation. This possibility is indirectly confirmed by Dr. Reinhardt, who worked at one of the secret scientific institutions toward the end of the 1930s.[20]

"I remember conferences during the war," says Dr. Reinhardt, "in which Navy officers participated. Regarding the project you're interested it, my memory tells me that it began much earlier than 1943—possibly in 1939 or 1940, when Einstein was engaged in working out the problems of theoretical physics presented to him by people who were interested in their military applications. The authors of that proposal were Einstein and Ladenburg…

"All summer and fall of 1939, Rudolph Ladenburg was at Princeton, working on experiments concerned with splitting the atom. I read that he discussed these problems with Einstein. In any case, I remember that this was in 1940, and the proposal that I associate with the ship project was allegedly the result of discussions between Ladenburg and Einstein about the use of magnetic fields for protection against mines and torpedoes…and Einstein himself wrote the proposal.

"Einstein and Ladenburg were always ahead when it came to submitting proposals, but in the presence of important personalities they preferred to keep a low profile. Von Neumann was a person of modest appearance, who knew how to draw those in power into his projects. So it was Neumann who talked

---

[20] "Reinhardt" is not the actual name of the informant, as investigators were required to preserve the anonymity of still-living participants in the incident.

to Albrecht, my boss, about this proposal, and one of them managed to actually get a green light from the research laboratory of the US Navy…

"Albrecht needed calculations for verifying the strength of the field and the practical likelihood of bending light to the extent of achieving the desired effect of a mirage. I swear to God, they had no idea what might result from this. If they did, the business would have ended right then…

"They discussed everything with Einstein, who even calculated the order of magnitude necessary for achieving the desired intensity, after which he spoke to von Neumann about what kind of setup would best demonstrate the possibilities of practical application. I don't remember exactly when the lab at the Office of Naval Research got involved, but Captain Parsons, one of the leading specialists in the US Navy, very frequently spoke with Albrecht – possibly about utilizing a ship…

"I seem to remember that, in one of the documents, I used the word 'deviation.' I also remember how, during one of the later discussions, I said that a ship could be made invisible with the aid of an ordinary veil of light smoke, and that I didn't understand why it was necessary to turn to such a complicated theoretical problem. In answer, Albrecht looked at me over the top of his glasses and said that I had an extraordinary talent for distracting people from the subject at hand…

"I attended at least one other conference, where this subject was on the agenda. We tried to ascertain the more obvious side effects that might result from such an experiment. The talk was about "boiling" water, about the ionization of surrounding air…but at that time, no one was able to consider the possibility of interspatial effects or the displacement of matter. In 1940, scientists attributed such things to the realm of science fiction. We wrote…that all of this is necessary to consider, and that in general, this whole business calls for utmost caution…"

Soon after this interview, Doctor Reinhardt suddenly died. It's possible that this was a random coincidence. But for some reason, certain participants in this whole episode, including Allende, wanted to keep a low profile even in the postwar years.

One is forced to consider the mystery of the Philadelphia experiment and the episode of James R. Wolf. He spent a lot of time verifying the testimony laid out in the letters of Allende. Wolf planned to publish the results of his research, but before his book was finished he abruptly disappeared.

Morris Jessup, soon after his visits to the Office of Naval Research, was found dead in his automobile, not far from his house. According to the official report, this was a suicide; suffering from depression, Jessup himself allegedly inserted a hose connected to the exhaust pipe into his car's interior. Death resulted from carbon monoxide poisoning. Some researchers, however, point to the fact that "a full inquiry was never made, which in itself is highly unusual for cases of suicide."

Doctor Manson Valentine was a friend of Jessup's, visiting him especially frequently during the last months before Jessup's tragic death. Valentine later expressed the suspicion that Jessup's depression developed because of emotional stress that he didn't want to publicize. "Perhaps he was allowed to die…"

2.8. Reverend Galiani's Suspicions

Why did the Korean Airlines passenger jet stray into Soviet airspace? A heated discussion erupted around this question, stumping politicians, journalists, and navigation technology specialists, in both East and West.

One of two possibilities: either KAL-007 intruded into Soviet airspace deliberately (a spy!), or it didn't. The latter implies that intrusion was the result of external circumstances—if the pilot did make an error, it was presumably by accident; or there were randomly occurring malfunctions in the navigational apparatus. This train of logic has given rise to two versions of events, Soviet and American. Both are deficient, as we'll ascertain later, but it seems a third one has not been proposed.

An analysis of the incident, to my view, must be prefaced by a fundamental observation: both versions largely formed according to specific mechanisms of psychological defense that eclipsed the unique circumstances mentioned above.

It was essential for Moscow to justify the destruction of 269 lives. This was possible only by being able to successfully show that the Soviet air defense services in the Far East had discovered the intruder in a timely manner and had exhausted all means of interception. Wasn't this why Marshall Ogarkov stated that one of the intercepting fighter jets had caught up with the airliner while still over Kamchatka, and tried to signal it to land? But if this had indeed been the case, the local command would have immediately contacted Moscow in order to coordinate further action.

The Korean airliner was signaled later, already in the sky over Sakhalin, by Major Osipovich, but KAL-007 didn't react to the signals. So then Moscow, without overintellectualizing the situation, declared: the airliner didn't react because it wanted to slip through. That means it was a spy. This was the genesis of the Soviet version. From this version, it follows that:

- The Korean pilots intentionally altered their course;
- American dispatchers noted on their radar screens that KAL-007 had deviated off course (but intentionally didn't react);
- Soviet radar registered the intruder in a timely fashion.

As may be seen, the Soviet version eliminates from the analysis of events all the unique circumstances mentioned above. Does this indicate a need to agree with the American version? NASA employee James Oberg answers this question in the affirmative. According to his opinion, Washington provides "an explanation at once the simplest and most likely." Let's check whether this is so.

Let us recall the prominent circumstances that led up to the tragedy. First:

- The radio direction locator at the Anchorage airport was undergoing repairs;
- There was a defect in on one of the onboard course pointers.

These two circumstances were obviously known by the captain of the airliner, and so can't be seen as "happenstances" leading to the deviation off course. The very fact that they have been ignored is witness to their insignificance.

Second: What happened during the first 49 minutes of flight? Much:

- The airliner deviates off course;
- The airliner fails to correct its course by the radio locator at Bethel.

Of course, these two circumstances allow us to propose both pilot error and malfunction of technical equipment. However, these possibilities seem unlikely even to proponents of the American version. The thing is, the dependability of several technical (as well as biological) systems is facilitated through a method of doubling.

Here, too: the plane has not one, but several navigational instruments, capable of substituting for or augmenting each other; the plane has not one, but two pilots, either of which may note a mistake made by the other. For these reasons alone, arguments about "pilot error" and "technical failure" appear weak. But what to do, when these are the only arguments allowing us to explain the incident by a fateful coming together of random circumstances, and spurn the accusation of espionage!

The Soviet version is strengthened by another unlikely circumstance. Namely: an on-duty monitor at a FAA radar station fails to notify the Korean pilots that the jet they're flying is off course on a track toward the Soviet border. If this coincidence were noted by some third-party observer, he would probably have suspected something wrong. He would have then and there remembered that the Korean pilots "for some reason" failed to correct their course as they flew over Bethel. His doubts would have increased even more, if he'd followed the further unfolding of events.

This outside observer would have to acknowledge that the dispatchers at other American monitoring stations, in whose field of vision the KAL-007 consecutively appeared, weren't in any way reacting to the fact that the airliner was inexorably approaching the borders of the USSR. Reasoning logically, he would have concluded that this was not by accident, that events are unfolding according to a previously established plan, that evidently there has been a prior agreement…

The first to draw attention to this suspicious detail was Mikhail Gorbachev. During the special session of the Politburo, convened the day following the incident, he noted that "the airliner was over our territory for a long time, and if it had strayed off course, the Americans could have notified the pilots. But they didn't do this."

Accusations of an organized spy maneuver rolled through the world press. On all sides there arose a question about the moral image of President Reagan's administration: as part of its political game, could it have put on the line the lives of 269 innocent people? The whole world waited for Washington's reply. But there were no concrete explanations. Hesitating with an answer, on the other hand, when Soviet diplomats were piling one "proof" on top of another, was dangerous. It's exactly here that the mechanisms of psychological defense came into play.

Moscow accuses: The airliner violated the borders of the USSR with objectives of espionage.

Washington: There were no objectives, the deviation from course was due to pilot error and malfunctioning of navigational instruments.

In this manner arose the American version. It's incapable of explaining why not a single one of the monitors at American radar stations, located along the Eastern coast of the Pacific Ocean, this region bristling with military facilities, reacted to the deviation of KAL-007 from its course. This circumstance raised suspicions even among Western observes.

The journal *The Nation* wrote: "It's impossible to believe the US didn't know that the plane had strayed off course. Why didn't we warn it?" The English journal *Defense Attaché* offers its own version: KAL-007 was operating in contact with the space shuttle *Challenger*. The basis for this argument—coincidence of timing: *Challenger* was also launched on August 31, and at an unusual time—at night, while the takeoff of Flight 007 was delayed at the Anchorage airport.

The journal lightly explains the consistent silence of American monitoring stations. These are the lines: "…As for the air traffic controllers monitoring the flight of the Korean airliner…they grossly violated their responsibilities, either not understanding that the plane was in danger, or failing to react when this danger was detected."[21]

Similar assertions played into the hands of Soviet propaganda, which immediately took them up and commented accordingly. Thus, the London TASS correspondent N. Pakhomov wrote that the routes of the *Challenger* and the Korean Boeing passenger jet "were in an ideal position with respect to each other for carrying out coordinated spying operations," and that "the space shuttle was engaged in eavesdropping on the Soviet air defense system."

Regarding the possible reasons for KAL-007's change of course, spokesman for the civil aviation pilots' association, John Meiser, attempts to assure the world community that in this case "it's possible to offer a convincing explanation…without seeking explanations of an otherworldly character." This allusion to otherworldly explanations wasn't made lightly. Indeed, a plane approaches the Soviet border, and the personnel of American monitoring stations don't react. If we allow that there is no connection between these circumstances, the thought of otherworldly forces involuntarily comes to mind.

James Oberg is fully aware of the weakness of American arguments. He understands that the coincidence in time of unlikely circumstances reduces the factor of randomness in the incident to a vanishingly small point. "In the

---

[21] In this case the speculation of journalists didn't go unpunished. Korean Airlines sued the journal *Defense Attaché* for slander. By court order, the publisher of the journal issued a public apology and, as compensation for moral damages, paid the aviation company a considerable sum of money.

process of reconstructing similar catastrophes connected to modern technology, we're always forced to confront 'unlikely' or 'improbable' occurrences that, nevertheless, occurred."

The word "unlikely" would have appeared fully correct in the above quoted statement even without quotation marks. Oberg, obviously, put it in quotation marks in order to justify the deployment here of the word—also in quotation marks—"improbable." Here the NASA employee is clearly dodging. Why he's dodging will become clear from a story borrowed from the book by the aforementioned mathematician Pólya.

"On one occasion in Naples, the Reverend Galiani saw a man from Basilicata, shaking three dice in a cup and taking bets that he could cast three sixes, and indeed getting three sixes. You'll say that such success is possible. However, the man from Basilicata was successful a second time, and the wager was repeated. He put the dice back in the cup three, four, five, times, and every time cast three sixes. 'Damnation!' shouted the Reverend Galiani. 'These dice are loaded with lead.' And so it was…"

In this case, the Reverend Galiani made a plausible conjecture. The thing is, the game of dice is based on randomness. If the game is conducted honestly, different numbers turn up when the dice are cast in succession. But when the game crossed the threshold into the realm of implausibility, which people view as supernatural, His Reverence lost his temper.

Moscow also accused Washington of playing a dishonest game. After all, they're trying to convince Moscow that at the same time that the pilots of the South Korean airliner are making their error, all the observers at American monitoring stations, military and civilian, are "erring" as well. Now it's clear why the NASA employee put the word "improbable" in quotation marks; if the quotation marks are removed from this word, a space opens up for otherworldly forces: the improbable has occurred!

2.9. "Einstein's Great Experiment"

Thomas Townsend Brown once got hold of an X-ray tube in order to find out what properties the rays emanating from it might possess. He hung the tube on a sensitive "balancer" and connected it to an electrical grid. As has happened more than once in the history of science, the resulting observation was not at all what the experimenter expected: every time the electric current was turned on, the tube rocked slightly. The X-rays had nothing to do with this, and Brown thought of another form of energy. In order to test his conjecture, he changed the specifications of the experiment.

A specially constructed apparatus was placed on a scale and connected to an electrical source of 100 kilovolts. Once the circuit was completed, the apparatus, depending on polarity, lost or gained nearly one percent of its weight. Then Thomas Brown was sure that he'd caught by the tail the up-to-then elusive force of gravitation, and named his apparatus the Gravitor. Although the invention was publicized in several newspapers, no one in the scientific world responded, possibly because at the time the author was a student in middle school.

Having completed his education in 1926, Brown works first in the Laboratory of the Office of Naval Research in Washington, in the field of physics, and in 1939 enters military service. He's a Navy lieutenant, a technical officer of magnetic and acoustic research in the Bureau of Ships. Later, having been promoted to Captain 2nd rank, Brown is transferred to a position with the Atlantic Fleet Radar School in Norfolk.

In December of 1943, at the insistence of physicians, he submits a request for discharge from the Navy. A brief bulletin in one of the American newspapers states that Thomas Brown, an outstanding physicist, was medically discharged following his participation in the Philadelphia experiment.

In the 1940s there begins a UFO "boom," with sightings indicating their possible extraterrestrial origin more frequently publicized. Brown throws himself wholeheartedly in researching this phenomenon. As the scientist becomes more familiar with the subject, he's increasingly convinced that the propulsion force of UFOs is electrogravitation.

A natural-born inventor, Brown conceives a daring experiment. In the first of a series of trials, he increases the lifting capacity of his Gravitor such that it can lift a weight exceeding its own. Then he gives the apparatus a discoid form and inserts a capacitor. And now when the disc, nearly a meter in diameter, is connected to an electric current, it rises from the floor and begins to fly in circles, buzzing and emitting a greenish glow.

In 1953, when Brown demonstrated his flying apparatus to the military, the invention was immediately classified, and everyone present signed an agreement of nondisclosure…

Isn't it strange, a discovery of utmost importance has been made (gravitation!), a discovery capable of achieving a technical revolution, a meteoric rise in scientific knowledge, yet to this day it remains without implementation?! This discovery is not only not mentioned in physics textbooks, it's not mentioned even in technical dictionaries and encyclopedias.

There can be only two explanations: either there were reasons for "mothballing" gravitational energy for a long time, or the invention had nothing to do with gravitation. This second reason was cited by several physicists, attributing the lifting capacity of Brown's apparatus to electric wind. But after all, apart from the capability of flight, the "disc" possessed other characteristics present in UFOs: it buzzed and emitted a greenish glow.

From the electromagnetic tape recording with Allende's narration:

"…So, you want to hear about Einstein's great experiment, yes? You know, I actually sank my arm up to the elbow in his extraordinary force field, which streamed counterclockwise around the small experimental ship—DE-137. I could feel the pressure of this force field on my arm, which I held in its humming, crushing flow.

"I could see how the air around the ship very gradually…became darker than the surrounding air…In a few minutes I could see a milky, greenish mist rising like a cloud. I think this was a mist made of elementary particles. I could see how, after this, DE-137 quickly became invisible to the human eye. If you tried to describe the sound accompanying this force field when it was swirling around DE-137…well, at first it was this buzzing sound, that quickly turned into a humming hiss…"

Continuation of the tape recording:

"…The field had a sheath around it, made of pure electricity. This current was so strong, it nearly knocked me off balance. If my entire body had been inside that field, it would probably have tossed me onto…the deck of my own ship. Fortunately, my entire body wasn't inside that force field when it reached maximum strength and density…my arm was pushed out by that field…The people from ONR still don't know what happened that time. They say that the field was 'overwound'…"

The similarity of physical effects inherent in UFOs and the physical effects which, according to Allende's descriptions, accompanied the Philadelphia experiment, are probably not accidental. Going by everything, it seems that the testimony laid out in Allende's letters is not a fabrication, that the results of Einstein's theoretical work laid the foundation of the Philadelphia experiment, while its initiators created (completely by accident) the physical characteristics of the field that is used in extraterrestrial technology.

It's also clear that Thomas Brown later utilized the results of the Philadelphia experiment in creating his model "flying saucers."

What would our life be like if, say, we didn't have electricity? This, of course, is hard to imagine. In addition to all the obvious, we must surmise that many technological discoveries, based on electricity, would have been delayed, and our civilization would today look quite different.

"I don't believe in the possibility of using atomic energy within the next one hundred years," Einstein once said.

The great physicist was mistaken: not a quarter century had gone by before there appeared atomic reactors, generating today over 20 percent of all the energy used by humanity. The question arises, how timely was the development of nuclear atomic energy? The legitimacy of such a question is obvious—if scientists hadn't unleashed atomic energy, there would not have been the tragedies of Hiroshima and Nagasaki, there would not be constant fear of their repetition.

Einstein thought that every method of destroying German fascism was good, and himself took part in the Manhattan Project.

"I clearly understood the terrible threat that achieving our project would entail for humanity," Einstein wrote. "But the fact that German physicists, working on the same problem, could gain success, forced me to take this step." When the scientist learned about the destruction of Hiroshima, his throat constricted so powerfully that he could only choke out the words, "Oh, woe!"

Subsequently, Einstein warned: "Our world stands before a crisis, the meaning of which is not yet clear to those who have been given the power to choose between good and evil. Atomic energy, released from its fetters, has changed everything; what hasn't changed is our way of thinking, and we move, defenseless, toward a new catastrophe…"

These words proved prophetic. In spite of international agreements about the nonproliferation of weapons of mass destruction, the threat of nuclear war hasn't disappeared. Scientists stand helpless before the problem of disposing the byproducts of atomic technology. The radioactive backdrop of the planet has increased catastrophically, as "acceptable" levels of radiation have been redefined more than once, after being found unacceptable.

God only knows what Chernobyl still has in store for us: inside the reactor, entombed in its concrete sarcophagus, an uncontrolled thermodynamic reaction sill cascades; there exists a real danger that someday there will be an explosion stronger than the previous one! In this sense the discovery of atomic energy may, of course, be considered premature. Many scientists compare atomic energy to a genie released from its bottle.

The research into the theory of a unified field did not, it would seem, harbor within itself any dangers; it was concerned primarily with the basis for the creation of the cosmos. Its discourses were in essence about cosmology, about the framework of our Universe. (And for Einstein it was a Universe in which vibrations of force fields are the cause of everything that goes on in it.)

Light speeds and paradoxes of time, gravitation and the bending of space— all this seemed purely academic, not having practical applications. But behind the abstract formulas there hid absolutely real forces, forces in reality cosmic. Perhaps it was then that Einstein beheld in his formulas a genie more frightening than the atomic one.

If we are to believe Allende, the genie was once released from its bottle to roam freely, but then they managed to chase it back into the bottle: the theory of a unified field was declared incomplete, and the Philadelphia experiment was declared a fabrication of journalists.

Meanwhile, the reality of events of great practical significance could have been confirmed by those mathematical equations from which issue the possibility of the very effect of teleportation. But do such equations exist in nature? In connection with that it would be interesting to find out, how far Einstein had gotten with the mathematical basis for his theory.

1927. Einstein publishes an article in a German physics journal, in which hesubstantiates the connection between the forces of gravity and electromagnetism; these were, in essence, preliminary results of the unified field theory.

1938. Einstein writes to his childhood friend, Solovine that he is currently working "with my young people on a very interesting theory."

1942. From Einstein's letters to Hans Mausam, a physician living in Israel: "I've become a solitary old man…But I'm working even more fanatically than before, and cherish a hope of solving my old problem of a unified physical field."

1944. To Mausam: "It may be that I'm still destined to find out whether I'm right to believe in my own equations…"

Einstein's biographer, Karl Zelig, once asked the physicist about his latest theoretical endeavors. "…As soon as the general theory of relativity was developed," Einstein replied in a letter, "the next problem immediately arose: this theory naturally led to a theory of a pure gravitational field…From that time, I attempted to first of all find a more natural relativistic synthesis of the law of gravity in the hope that such a generalized law would serve as the common field theory…To my deep satisfaction, I succeeded in obtaining the necessary equations…"

These lines were written on March 25, 1953. But a few months later (September 14), Einstein, as if catching himself, writes an addition: "…The mathematical foundation of the theory is irrefutable. However, the question of its physical meaning is not completely worked out, since for comparison with an experiment it's imperative to find numerical solutions to the field equations, which isn't always successful. This situation may possibly last a long time…There's not much hope that I'll achieve success in the few years still remaining for me to be able to work."

Reading into this second letter from Einstein, it's hard to shake the feeling that the scientist regretted having rushed to announce his obtaining of the "necessary equations." Necessary—for what? This question may be precisely answered with a relevant experiment. Einstein probably hoped for just such an experiment: "It may be that I'm still destined to find out whether I'm right to believe in my own equations."

If the equations for the theory of a unified field already existed in 1943, then doubtless they served as the basis for the Philadelphia experiment. In that case Allende's assertion, that the work underlying the theoretical foundation of such a dangerous experiment was declared unfinished, rings absolutely true.

The declaration was made, most probably, on the recommendation or even insistence of the architect of the theory himself, apparently judging the experiment premature. Einstein was tormented by the consciousness of his contribution (though indirect) to the creation of the atomic bomb, and it's natural to think that he didn't want to add another burden of sin to his soul.

Could Einstein have, for humanitarian reasons, destroyed the fruits of his half-century labor? Indeed, he could, those people who knew him personally wouldn't hesitate to answer. The possibility can't be excluded, however, that the handwritten pages on which the great physicist worked out his field equations are not to this day preserved under a "Classified" stamp in an armored safe of one of the military institutions of the USA.

2.10. A Vicious Circle

(Scrolling through Russian newspapers)

A scholarly member of the political science faculty at Oxford University, R. U. Johnson, devoted several years to studying the circumstances of the Sakhalin incident. His major focus was on possible causes that may have led the South Korean airliner to stray off course.

…And so, at the moment of its destruction from a fired missile, KAL-007 was over 500 kilometers from its assigned route. How could that have happened? Johnson enumerates the possible reasons:

- "– Both the INS and the autopilot went out of order, but the crew didn't notice the yellow warning light on the control panel that signaled this[22]…
- A false flight program was intentionally inserted into the plane's INS in Anchorage, with the intent of flying over Soviet territory. This could not have been accomplished without the crew's knowledge…
- A mistake was made in programming the INS in Anchorage, the autopilot aimed at the radio beacon in Bethel malfunctioned, and the crew didn't notice…There is no other alternative…"

It was precisely this last, third alternative that US President Ronald Reagan latched onto. Soon after the tragedy he announced that the crew of Flight 007 most likely incorrectly entered the data into the onboard autopilot system, causing the jet to go off course.

But 10 years later the members of the investigative commission, which included experts from the US, Russia, South Korea, and Japan, agreed that the

---

[22] Here, INS stands for Inertial Navigational System.

likelihood of incorrect flight data entry into the computer was essentially excluded. Regarding the reliability of the navigational system itself, they write, "[it] constitutes a wonder of modern technology, consisting of three computers, such that even if two of them fail, the system retains its functionality.

The initial position of the plane is entered into it, then the point of destination and intermediate control points…This system not only guides the airliner with incredible accuracy (deviation may be no more than 1 mile over a 5,000-mile route). It tracks and corrects deviations of speed and height…and many other parameters.

In order to prevent human error in programming, the program is entered into the computer complete, recorded on cassettes, which are specially packaged together with the flight plan. But after that comes a mandatory checking procedure, when the flight engineer "plays back" the cassette on his computer, while the pilot and co-pilot monitor the conformity of the data with the flight plan on their own computers."

All this the crew of KAL-007 performed at the Anchorage airport, not long before takeoff. Even if, however, the navigational complex had for some reason broken down, the regular magnetic compass in the pilots' cabin (independent of all the other systems) would have shown the deviation…

"To think that 007 accidentally strayed off course at Bethel—as deemed by some analysts—is to believe that Captain Chan and his officers were guilty of extraordinary sloppiness and inattention."

Nevertheless, over the course of the following years, experts never stopped talking about the possibility of error in the navigational apparatus. "Traffic control services of several countries, even after the Boeing perished, didn't suspect that it had strayed off course," wrote one of the Russian newspapers toward the end of 1992. "And the accepted explanation for this is that the crew, relying on the indications of faulty instruments, was giving ground controllers distorted data about its position…"

However, if the version about instrument failure is accepted, there's the disconcerting fact that, out of all the countless variations of possible "distorted data," the apparatus gave out precisely those that were in complete accord with the program put into the computers—after all, the Korean pilots, all the way up to the catastrophe, didn't display any concern. Is such an improbable coincidence possible?

Obviously, the English political scientist doesn't believe in the randomness of such coincidences. But if the navigational equipment of the Boeing was functioning correctly, and the crew made no mistakes, what is left? Only one thing is left: to accept the espionage version. To the protests of his adversaries, Johnson counters with Occam's principle: "there's no need to complicate arguments any more than necessary."

"Nevertheless," adds Johnson, "it behooves us to consider all oddities and possibilities in the manner of Sherlock Holmes."

The facts that disrupted the seemingly coherent system of proof upon which the espionage version was based were publicized after the dissolution of the USSR. In 1993 a government commission was convened in Moscow by order of Russian President Boris Yeltsin, called to investigate the circumstances of the Sakhalin incident, and to consider these "fresh" data.

Soon afterwards, the head of the commission, Sergei Filatov, announced that no espionage equipment had been discovered amid the wreckage of the plane; there was no evidence that reconnaissance was the purpose of the flight. (These facts, obtained by Soviet divers from the bottom of the Sea of Okhotsk, had been concealed for the duration of the following decade.)

Additional investigations were also carried out by the International Civil Aviation Organization (ICAO). Their conclusion: the instruments on the Boeing were in correct working order.

A puzzle had developed, worthy indeed of the famous English detective's attention: a plane with working navigational equipment ventures deeper and deeper into Soviet airspace, but its crew isn't pursuing a reconnaissance mission. A glaring absurdity! Logic compelled a return to the idea that the pilots had been negligent, a version that, for reasons of its high unlikelihood, had been discarded by most researchers.

Izvestia [News]
June 25, 1993
"The crew relied entirely on the autopilot, engaged right after KAL-007 lifted off into the air above Anchorage, and disengaged only five-plus hours later, when the Boeing had already been crippled by Osipovich."

Trud [Labor]
June 31, 1993

"What could have induced the pilots of the South Korean Boeing, over the course of five hours, never to doubt their position? And this despite the numerous indications of instruments that should have warned the pilots…A pilot can't – no way! – not notice that he's drifting to the side…Judging by official data, the crew conducted itself negligently in view of the situation. Why? The motives are incomprehensible…"

Komsomol'skaya Pravda [Komsomol Truth]
August 31, 1993

"The ICAO's investigation has been completed. The published conclusions of the commission indicate: the instruments on the Boeing were in correct working order. The crew of the jetliner, by not paying attention to the deviation off course, at minimum was guilty of negligence…"

The puzzle, however, isn't solved: by concentrating on the criminal negligence of the crew, reviewers of the incident inadvertently revive the question: why did the airliner, whose course was set and controlled by computers called a miracle of modern technology, deviate from its assigned route? This contradiction was addressed by a different newspaper.

Rossiiskaya Gazeta [Russian Newspaper]
September 22, 1993

"Of course, it's unlikely that the crew would have been carrying out a reconnaissance mission—the plane didn't have the necessary equipment. But the fact that the flight, from the start, proceeded on an unassigned itinerary and was motivated by unknown reasons—this fact has been completely ignored, both by the ICAO, and the presidential commission…"

True, physical evidence of a reconnaissance mission by the flight was lacking. But what other evidence is needed for this? After all, according to experts, the pilots of KAL-007 couldn't ("no way!") be unaware of the deviation. And, in opposition to the presidential commission and the Russian press, accusations of espionage were sounded once more.

Trud

September 11, 1993

"The pilot of the jet, Chan Ben In, had flown planes along this same itinerary 27 times, and, by the way, only five days before his demise; on August 26 he flawlessly completed the very same route. What talk can there be of negligence?..."

The cited excerpts from Russian reporters graphically demonstrate the way they kept going around a vicious circle, turning again and again to old versions that have never been proved.[23] But, devil take it, there has to be some sort of explanation here!

---

[23] The same cacophony concerning the reasons for KAL-007's deviation off course reigned in the pages of the Western press.

2.11. Dubious Arguments of an American Expert

In 1982 a Soviet Ilyushin airliner deviated from its assigned route by 150 kilometers and invaded Switzerland's airspace. Two Swiss air force fighter jets forced the intruder to land at the airport in Zurich. No physical evidence of espionage was found on the Soviet airliner. But since the Ilyushin was flying over the region of St. Gotthard, where military maneuvers were taking place at just that time, Geneva sent Moscow a note of protest. The director of Aeroflot in Switzerland, Leonid Barabanov, immediately departed for Moscow "under threat of being accused of espionage activity".

A year after the Swiss incident, the airspace of the USSR was invaded by the South Korean airliner KAL-007. Soviet diplomats accused the USA of espionage. One of their arguments: the jetliner flew over a secret strategic zone on Kamchatka. The ICAO considered this argument meaningless, since the entire Soviet Far East was one solid strategic zone. But, after all, crowded Western Europe is completely covered with zones containing strategic NATO installations.

As may be seen, in analyzing one or another incident of a plane violating the air boundary of a different sovereign government, emphasis lies overwhelmingly on the political aspect. This happens, apparently, because it's uncommonly hard to refute (and, incidentally, to prove) that this or that intrusion was premeditated.

Soon after the tragedy of KAL-007, international commissions looked into a series of incidents in which planes went off course. Also, under discussion was the general question: how often do these events take place? Here opinion was divided. The Soviet side insists that incidents of unintentional deviation of planes from their course are exceedingly rare. American experts say the opposite: the deviation of a plane from its course as the result of human error or malfunction of navigational equipment is not such a rare thing.

NASA space engineer James Oberg attempts to support the American point of view by means of analogy.

He writes: "An analysis of aircraft deviation from course over the northern portion of the Atlantic over the past two years revealed several incidents of this nature which, had they occurred on the route of [Kal-007] in the northern portion of the Pacific, would have inevitably led to an invasion of Soviet airspace…"

Stop! Why did the NASA engineer specifically choose the northern Atlantic for his example, and not another region of the planet? And, while he was at it, why didn't he indicate concrete reasons for these deviations from course? Why, because it's precisely in the northern Atlantic, especially in the region of the Bermuda Triangle, that planes most often go off course for unknown reasons. In 1983 alone, 55 such deviations were registered. Isn't that too many? It's a lot, of course. One even wonders at the unreliability of Western navigational technology.

"It's interesting to note," Oberg writes further, "that Aeroflot airliners on international routes have recently made mistakes four times…"

The NASA engineer apparently wants to say that Russian pilots also often make mistakes. The frequency of incidents, as is known, is determined by their quantity over a specific period of time. Four mistakes made "recently." Over what period of time? It turns out that what Oberg has in view are two incidents in 1982, and two in 1983. Now we may ask if this is many incidents, or few? Let's check.

Aeroflot is one of the world's major airline companies. The extent of its routes consists of over 200,000 miles, connecting 80 countries. Therefore, with equal dependability of navigational equipment, and equally professional pilot training, Aeroflot planes should naturally stray off course more often than the planes of Western airline companies. But, four incidents over a period of two years, and this in view of the enormous extent and passenger load of the airline—that is really not very much.

Let's compare: four "Russian deviations" over two years, as opposed to 55 "Western deviations" (in the north Atlantic) in 1983 alone!

If we agree with Oberg's arguments, the specified correspondence should testify: Russian navigational technology is the best in the world, while Russian pilots make far fewer mistakes in flight than pilots of Western countries. However, the dubiousness of such a conclusion leads one to think of a different, more likely explanation: the reason planes stray off course over the north Atlantic lies in some sort of external physical factor.[24]

…The crew of KAL-007 was supposed to run an inspection of the course as the airliner approached Japan. This inspection is performed with the aid of atmospheric radar that "probes" the surface of the Earth. The images displayed on the monitor screen (contours of shorelines, curves of rivers, mountain chains) are compared with a geographical map. Why did the Korean pilots, who had flown this route more than once, make a mistake on this occasion?

James Oberg undertakes an explanation of this circumstance, too: "What would the pilots of the Korean airliner have seen on the radar screen?" he asks. And answers: "They expected to see the coastal outline of the northern part of Honshu…And they did see a coast. But by unhappy coincidence this totally identical configuration was that of a different region. The shoreline was the shoreline of Sakhalin."

---

[24] It's surprising that representative committees investigating possible reasons for KAL-007's change of course did not draw parallels with other incidents of intrusion into Soviet airspace in the same region.

On June 30, 1968, an American Seaboard World Airlines jet transporting troops to South Vietnam intruded into Soviet airspace in the Kuril Islands region.

On April 20, 1978, jetliner KAL-902 embarked for Seoul out of Anchorage. Over the Bering Sea the jet changed course and strayed deeper and deeper into Soviet airspace. The fighter jet sent to intercept it "circled the mysterious guest for 10 minutes, emerging 50 – 100 meters in front of the pilot's cabin." As the result of a fired missile, two passengers were killed, and 13 injured. On the ground, the Korean pilots were asked why they hadn't been able to orient themselves, when the sun shone into the cabin from a totally different direction? There was no coherent answer.

On September 29, 1984, fighter jets of the Norwegian air force intercepted an American South Pacific Island Airways plane heading for the Soviet border. The jet, flying from Anchorage to Amsterdam, had strayed 900 kilometers from its assigned course.

On October 11, 1985, a Japanese Airlines Boeing 747 came within a hair's breadth of the Soviet border. Thanks to timely measures taken by rescue services, a potential tragedy was averted.

The NASA engineer speaks of a fateful coincidence. And I ask again and again: aren't there too many coincidences in one incident to speak of randomness?

The basis of English political scientist Johnson's espionage version was the observation that, even if the navigational complex went out of order, deviation from course shown by the independent-of-other-systems magnetic compass couldn't be missed by the Korean pilots. The argument would seem to be irrefutable. But only at first glance. First of all, the compass needle may deviate by several degrees in zones of magnetic disturbance, and when the plane traverses an electrical field it begins to "dance." Secondly…

It happened in 1955. A DC-6 passenger airliner was on its way to Bahrain. It was a night time flight. The passengers slumbered in plush chairs, lulled by the even humming of the engines. In the cabin, only the stewardesses were awake.

Suddenly one of them noticed that the moon, which had recently vanished from the cabin window, was once again in her line of sight. The young woman thought at first that she was seeing things. But a little while later she saw the moon again, in the same cabin window. This happened again and again. For some reason, the airliner was making big circles over the Mediterranean. The stewardess headed for the pilots' cabin to find out what was going on. Here she witnessed a striking scene: the entire crew was fast asleep.

An analogous incident also took place in 1955. "The stewardess, noticing that the plane was in a steep dive, decided to find out what was going on, and discovered a sleeping crew. The pilot, suddenly jerked awake by the woman's screams, had just enough time to disengage the autopilot and pull the rudder toward himself."

Incidents where pilots fall asleep on flight routes are studied by psychologists and physiologists. These incidents receive explanations. Here is one of them. While the plane is on autopilot, the crew has far less work than the flight attendants serving passengers in the cabin; for most of the way, one might say they have almost nothing to do. The monotony of the flight, the lulling drone of the engines, the awareness that the plane's course is maintained by intelligent apparatus, all serve to dull alertness and produce relaxation.

Another explanation: on long flights the plane crosses time zones, which leads to a disruption of the biological clock responsible for regulating rhythms of wakefulness. The pilots' psychic activity may also be inhibited by magnetic disturbances, if the plane happens to be in a zone of their activity.

Biophysicists figure that evolution had to have worked out a mechanism for protecting organisms from the effect of abrupt fluctuations of the magnetic field, sporadically occurring in the atmosphere. It's possible that one of these defense mechanisms is sleep—no wonder Russian physiologist I. P. Pavlov called sleep "a protective deceleration."

Evolution invented and embodied in organisms a multitude of clever mechanisms, to stand against harmful factors in the external environment. It just couldn't predict that its youngest child would think up airplanes…

The pilot of the Soviet jet fighter, Major Gennady Osipovich, followed the Korean airliner in total for 26 minutes. During this time, he not only called out the "intruder" by radio, but four (!) times he opened warning fire and "came right up to [the airliner]" in order to draw attention to himself.

But the Korean crew never noticed any of it. It looks as though the western region of the Pacific, as well as adjacent sections of land, are also sporadically covered by "a giant camouflaging electronic net." Not as often, however, as in the northwest Atlantic.

2.12. Within the "Zone of Invisibility"

From the interview of a correspondent of the newspaper *Izvestiya* with Lieutenant Colonel Gennady Osipovich, who shot down the passenger airliner KAL-007:

"…Eight minutes of flight had gone by. Suddenly the guidance navigator transmits: 'The target is in front of you, the intruder jet…it's heading your way.' I reverse my course. And, after getting height adjustments, I go for the intruder. The weather was fair. Through thin clouds I soon saw the intruder jet…

"What were you thinking at that moment?"

"I wasn't thinking anything. There was a hazard! What is a fighter pilot? Kind of like a sheepdog that's always being sicced on a stranger. I could see it ahead of me—the stranger. After all, I'm not a traffic inspector who can stop a violator and demand documents! I followed it, to cross its path. The first thing I needed to do was to force it to land. And if it didn't comply, neutralize it by any means. I just couldn't think about anything else. All the rest that I heard later—it's all irrelevant stuff. Nothing else.

"So then, coming up on him, **I locked onto him in my radar-controlled aimer. Right away the socket heads on the rockets lit up.** I announced: 'The target is in my sights. I'm going after it. What do I do?'

"But the guidance navigator suddenly started asking me about stuff: course, height of target…Though it should all have been the other way round! And only later it became clear: we were both moving in a zone of invisibility, the existence of which no one had even guessed…

"'For some time, we couldn't see you or him,' the navigator told me later, on the ground.

"Finally, we were approaching Sakhalin. And then the navigator gives the command: 'The target has violated the government border. Eliminate the target...'"

Major Osipovich shot two rockets into the Korean Boeing. As he stated later, one of them sheared off half the left wing. The second rocket, according to experts, must have damaged one of the four engines. Osipovich saw the rocket explosions, and this was enough to inform the control station: "The target is eliminated!"

And who could doubt it, since the target was locked onto by the radio-controlled aimer, and the rockets were equipped with self-guiding instruments! However, the pilots of the other Soviet fighter jets told the control station that the intruder jet was still flying.

From the transcript of conversations between Soviet pilots and ground control:

"Titovnin, what do you have?"

"Nothing as yet."

"What's the matter? It was locked onto, why wasn't it shot down?"

"Titovnin, what about it, tell us. Did they lose the target?"

"They lost the target, comrade colonel, over Moneron Island...after the launch, the target made a correct turn over Moneron..."

"Finish it off, finish it off. Go on, bring the MIG-23 closer..."

However, finishing it off proved unnecessary: KAL-007 vanished from the radar screens. Subsequent analysis of the tapes in the "black boxes" led experts to a truly sensational conclusion. Contrary to the announcement made by Osipovich, from on board KAL-007, even after the attack, radio signals continued to be sent by the antenna located precisely on the tip of the left wing.

Contrary to the thinking of experts, not a single one of the four engines was damaged; twice, the flight engineer commented—and this is clearly heard on the magnetic tape recording—that the engines are functioning normally. Moreover, after being hit by the rockets, KAL-007 didn't start to fall, but instead rose over a kilometer.

Only after that did it begin a descent, though not falling, but gliding in spirals. At least nine minutes after the attack, KAL-007 still appeared on air defense radar screens, at a height of 5,000 meters. At this point, for unknown reasons, both "black boxes" ceased to function.

…Albert Einstein cherished a dream that the theory of a unified physical field, on which he worked especially intensively during the last years of his life, would prove the reciprocity of forces in nature. As Einstein's biographers write, the scientist didn't complete his endeavor, but his idea remains alive, still exciting the minds of theoretical physicists today. According to their representation, the unified physical field may be broken down into three vectors: electrical, magnetic, and gravitational (which is the vector of inertia). In specific circumstances, one vector may be replaced by another.

This hypothesis has been embraced by ufologists. Probably, they conjecture, passengers of UFOs are capable in this way of altering the conditions in which one physical vector is replaced by another. If, for example, during a sharp change in flight direction the gravitational vector is compensated for by the electrical vector, then the passengers of the UFO would experience no physical overload, not even sensing the turn.

In conditions of zero gravity, the device could start its motion practically at any speed, and also come to a sudden halt. Some researchers connect changes in the brightness of an object to similar fluctuations of physical fields. Thus, in the moment that the gravitational vector is compensated for by the electrical vector, the electrical potential of the body sharply increases; seen from the side, this would look like a bright flash…

Much was said about light effects in connection with the warning shots fired at the Korean Boeing by Osipovich.

The Soviet pilot later talked about these warning shots during a television interview: "I sent four rounds of tracers right under the nose of the Boeing. They can be seen over many kilometers, and of course they had to have seen them…"

The circumstance of the crew of KAL-007 not even noticing the tracer rounds is hard to explain. American Senator John Glenn, a former military pilot, says: "If there were tracers, it's impossible to explain why they didn't notice them!"

The witnesses to these minutes were fishermen on the Japanese vessel *Sidori Maru* 58. Crew members said that around 18:00 hours they heard two or three rolls of thunder and saw a flash in the sky. If the fishermen were correct about the time, then it was a bit early for the hitting of the target, but if they could see…the tracer rounds at a distance of 6 kilometers, it's even harder to explain why they weren't noticed by the crew of the Boeing.

From the interview by a correspondent of *Izvestiya* with Lieutenant Colonel Osipovich:

"...I engaged the afterburner...the rocket heads started blinking. I told the ground—the target is locked onto. And suddenly in my headset: 'Stop elimination. Go to the height of the intruder and force it to land.' And I'm already coming up to the intruder from below...Fire warning rounds—I hear from the ground. I fired four rounds, used up over 200 shells. And for what! I've got the armor-piercing kind, not the flares. It's not likely that anyone can see them anyway..."

"But our newspapers reported that you fired warning shots using flares – glowing tracer shells..."

"That isn't true. I just didn't have those kinds of shells. So I fired the armor-piercing kind..."

It turns out that in the television interview, conducted long after the advent of Gorbachev's Glasnost', Osipovich wasn't being honest. Simply put, he told a lie. He lied, apparently, on orders from above. With regard to the flash of light seen by the Japanese fishermen, I invite the reader himself to guess its origin. Moreover, one should consider: "it was early for the hitting of the target."

2.13. Face to Face with a Mystery

It's hard to say who was first among science fiction writers to play on the idea of "transmitting" a person over a distance by means of a beam of light. But an indication of the possibility of realizing such an idea may be found in a book written long before the advent of science fiction. What I have in mind is the Bible.

The reader is of course familiar with the story of the miraculous resurrection of Jesus Christ, and those who enjoy science fiction may remember storylines about how Jesus was sent to Earth with a mission from another constellation. In accordance with Hebrew custom, the body of Jesus was wrapped in a winding-sheet of linen cloth and laid in rock cave. The entrance to the cave was then blocked by a huge stone.

"…So they went, and made the sepulchre sure, sealing the stone, and setting a watch. In the end of the sabbath, as it began to dawn toward the first day of the week…behold, there was a great earthquake: for the angel of the Lord descended from heaven, and came and rolled back the stone from the door…his countenance was like lightning, and his raiment white as snow: and for fear of him the keepers did shake, and became as dead men. And the angel answered and said unto the women, Fear not ye: for I know that ye seek Jesus, which was crucified. He is not here: for he is risen…"[25]

This story might be considered a fiction, if not for the shroud which, according to testimony from the apostles, was left behind in the cave after the disappearance of Jesus' body. On the shroud it's plain to see the contours of a human body from the front and back. The hands and feet have been pierced, and there's a wound between the ribs.

---

[25] Matthew 27:66, – 28:5 (KJV).

The face is symmetrical, with a high forehead on which may be seen traces of the "crown of thorns"; long hair, large eyes, whiskers, a beard. The image resembles a negative made by photographic method.

When the Shroud was exhibited in Turin in 1978, some American specialists in radiation technology were allowed to inspect it. The results of their inspection, causing great consternation in the ranks of atheists, showed: "The shroud was wrapped around the body of Jesus Christ, whose image was imprinted on it by means of a flash of radiation…emanating from every part of the body."

Conclusion: the negative imprint on the Shroud does not resemble the work of an artist, but represents the result of a photochemical effect resulting from an intense flash. The body of Jesus dematerialized, while wrapped in the Shroud!

If one remembers that later on Jesus came to his apostles in the flesh, while several vanished sailors from the destroyer *Eldridge* came back from "another world" where they communicated with strange creatures, then it becomes clear: we're talking about the teleportation of people—into a parallel universe and back. These events, apparently, were affected by the activity of a hyperfield, whose fluctuation contained an encoding signal.

As happens in similar cases, there followed publications with assertions that the image on the shroud was an expertly executed fake. We can't, of course, exclude this possibility, but to people who think discussions of teleportation are absurd, I recommend heeding the words of Norbert Wiener, the "father" of cybernetics.

"We represent in ourselves not a substance that is preserved, but a form of construction that perpetuates itself," wrote the American scientist. "A form of structure represents in itself a signal, and it may be transmitted in the form of a signal…It would be interesting and instructive to examine what would happen if we were able to transmit the entire form of the structure of the human body…in a manner such that the hypothetical receiving apparatus could reconstitute these signals into the corresponding material, capable in the form of a body and brain to continue the processes of life…"

And further: "There is no difference between the types of transmission we may employ for the sending of telegrams from one country to another, and the types of transmission that…are capable of transmitting live organisms, such as human beings. In that case…the idea of the possibility of travel with the aid of the telegraph…does not appear absurd, however far it may be from realization."

As I was finishing this book, a newspaper article came my way, reporting that all five planes of Flight 19 "were discovered at a depth of 250 meters, only 15 kilometers from Ft. Lauderdale. They were accidentally discovered by the vessel *Deep Sea*, equipped with controllable underwater cameras."

Four of the planes were well preserved, while the fifth had broken in half…The number "28" was clearly visible on the broken plane—the number of the machine flown by Lieutenant Taylor. On another plane were the letters "FL," with which combat machines out of the base at Fort Lauderdale are marked…Where did the crew of the Avenger planes go?

Underwater photos show completely intact glass domes on the cabins, but no trace of people. And if they managed to escape with parachutes, then why weren't they observed on the water by the search planes?

From an interview by a correspondent of the newspaper *Trud* with a diver who participated in the search operation at the place where the Boeing 007 went down:

"…What did you find?"

"Components of the Boeing, articles, money, all kinds of minutia, and suitcases, lots of them…"

"What conclusions do you draw from this?"

"Let the commissions draw the conclusions. But the fact that things were found, but no traces of people, you'll agree is strange."

"Do you mean to say that if there were passengers on the Boeing 747, their remains would have been observable on the ocean bottom?"

"Without a doubt…Judge for yourselves. The planes of the Soviet air force approached the Boeing close up, tried to connect with the crew, there were, if official material is to be believed, four rounds of warning shots with tracers. And no reaction…"

"You wish to say that the Boeing was pilotless?"

"Definitely. And without passengers. This version is upheld by all my fellow divers…"

"A decade has passed, and it's hard to imagine that such a version hasn't received factual support."

"But no one is challenging it today. And then, we all submitted signed nondisclosure agreements about the mystery to the proper authorities. One of my best friends spent a whole week on the bottom—and no sign of the deceased."

"Let's assume you simply didn't find them…"

"Impossible. I know what I'm talking about. At the start of the 80s, a Yakovlev combat plane went down in the Cam Ranh region. Over the course of a month, descending to the bottom of the supposed area where it fell, we found the remains of the pilot; after all, we're professionals. Our profession is to salvage. Sometimes we find even tiny parts of things that sank decades ago. Here there were 269 people—and nobody. Unbelievable."

"Have you come up with a reason for yourself, why the true details of this tragic event have been hushed up?"

"We thought about it for a long time, and agreed on one version: there's collusion between interested powers on the American and South Korean sides, and we don't exclude the participation of our own intelligence services…"

N. Fedoseyev, retired Captain 3rd rank:

"On the night of August 31, 1983, while guarding the state border in the region of Moneron Island (30 miles from Nevel'sk), I received an order from unit headquarters to go to an area where there was a presumed accident with a foreign plane. At 8:00 we approached the indicated place, where there was already another ship form our unit. After verifying with headquarters that our task was to fish out everything that was floating…we started to carry it out. And there was lots of stuff floating: the aircraft skin with a plug, lots of children's, men's, and women's clothing, documents.

"The weather was excellent: sunshine, calm. All the items were floating within a mile-and-a-half radius. On the first of September everything we'd collected was turned over to the unit…Coming down to the beach, I looked over the contents of the **containers**. There were two of them, five tons apiece. I didn't see any bodies, or body parts…It's clear that the Boeing 747 didn't have any passengers…"

From an interview by a correspondent of the newspaper *Trud* with a French specialist in airline disaster investigations, M. Philippe Henneken:

"As far back as 1983, we were asked to help investigate the auto recording of the Boeing. We were given the magnetic tapes in January of last year [1993] at the European headquarters of the ICAO, and we worked on them in the presence of representatives from four interested parties—Russia, Japan, South Korea, and the USA."

"That must have been arduous work. Are there any results yet?"

"There were two auto recordings. One of them is parametric…which aids in recording the parameters of the flight. We had the voice recorder, which had the communications between the pilots and the ground…"

"There are many versions regarding the tragedy with the South Korean airliner. One of them posits that the recording on the voice tape was made previously, and the plane itself had just the hardware…"

"We're dealing with the original. From a technical point of view, there is no doubt that the recording was made during flight."

"But extraordinary things really did occur after the Boeing was shot down. The search party never did find any remains of people on the bottom. Plane fragments—yes, a heap of suitcases—yes, but no people."

"It's hard for me to answer that question. Especially since we weren't looking into that aspect. Personally, I have no doubts that the plane was performing a commercial passenger flight, that there was a navigational mistake leading to the well-known tragic consequences, and also that there were people on board.

"Listening to the voice recording, you're convinced that there really is a crew on board. It's absolutely true that the pilot's phrase, "What's going on?" occurs on tape at the exact moment when the rocket pierced the airliner. You can clearly hear it on the tape. I can't say more at the moment…"

R. Mikhailov, doctoral candidate in technical science:

"…Feeding Soviet citizens a load of primitive garbage, official parties investigating the incident spouted nonsense about 'fast currents' at the site where the plane fell. So the upshot is, even if currents unknown to geographical science carried away all the deceased passengers, at least the crew should have been preserved, being in a hermetically sealed cockpit separate from the plane's cabin!"

V. Malyutin, former USSR test pilot:

"The story of the destruction of the passenger airliner…with 269 passengers on board, somehow doesn't fit together well in the mind. What is this, a sophisticated provocation, or plain old vandalism? There are plenty of stories, but little reliable documentary evidence, and what there is, isn't convincing. So, what is this destroyed plane, a "pilotless tool," or an actual passenger airliner with a crew and passengers? As a specialist, I tend to favor the first alternative—a Boeing 747 equipped with reconnaissance apparatus, and minus passengers…

"You'll agree that the crew of the passenger airliner conducted itself very strangely under attack (if there was indeed a crew on board). It's night time, chains of glowing tracer shells are going by right under their nose. The first logical response of a pilot is to maneuver out of the line of fire, and at the same time…to give a signal to the effect of 'I see you, and I'll obey your commands.' That is, save your passengers and yourself. And what does this one do? Nothing, he's like a suicide bomber, he doesn't react to anything, doesn't execute any maneuvers.

"How can you help calling him an automaton? And the thing is, we know who the crew was, headed by an experienced pilot, and a military one at that, Chan Ben In. What for would he turn into a kamikaze? You have to be crazy to act that way.

"It's very doubtful. And, as a specialist, I'm not convinced by the conversations with the ground, or by the conclusions of the ICAO commission, that considered its work accomplished without establishing any motives for the behavior of the crew…I'm convinced: it was a meticulously planned and executed act of provocation. And its goal was to show the entire world the inhumanity of the Soviet military and leadership…"

V. I. Nagorny, reserve Colonel, commander of security services of the Russian Air Force:

"My attitude toward the 'pilotless' version is this. The manufacture of a convincing duplicate of a Boeing 747 (the world's largest passenger jet), in a one-off model, for achieving a private reconnaissance mission unlikely to have any strategic importance, seems like an excessively costly gratification. Even for the USA.

"In my "combat" past I had to carry out over 50 real shootings of radio-controlled targets with rockets and cannon. So I understand these devices very well, the great technical difficulties in producing and operating them. By the way, the only reconnaissance goal that a flight like that could have had on that particular night would be the discovery of new frequencies, on which air defense radar transmissions were conducted, according to plan…

"Such changes of frequency are conducted regularly, and aren't a secret to anyone, or at least, not to specialists: new frequencies are discovered by an opponent through electronic means a lot more simply, cheaply, and with fewer dramatic effects than on that tragic September night…But the question, were there passengers or not, must be investigated not by aviators, but by specialists of a different field. It's surprising that the general public still doesn't know anything about this…"

Yu. Maleyev, doctor of jurisprudence, president of the independent institute of air law:

"…Why no bodies of passengers were discovered at the place of destruction, such a thing has never been met with in the history of aviation. Why, in 1986, for instance, on the 23rd of June, when an Air India jet blew up over the Atlantic, the same kind of Boeing 747they found over 100 bodies at depths two and a half times the deepest parts of the Sea of Okhotsk, even though the search operations weren't as intensive or prolonged…"

V. Zakharchenko, leader of a team of divers:

"…Well, we found trousers with holes in the knees, a belt also ripped out, but everything else was intact. What does this tell us? A person was probably sitting in these trousers…You know, all kinds of talk went on in this expedition: like there weren't any people there, on that plane, that this was all a fake. Generally speaking, I also had a similar opinion at the start. Almost no traces that there had been people there, except for personal effects. But the things were there, after all!

"Later, after we returned to Murmansk, we started reading the papers—we were especially curious to see what they were writing. I thought then—you can't fake the death of that many people…organize their relatives to grieve for them—in Korea, in Thailand, in the USA, in Taiwan…Well, maybe you could fake two or three, but not two hundred plus…"

The proposition that the passengers' corpses were carried away by underwater currents doesn't stand up to criticism. As the investigation showed, after being attacked the jet survived for at least nine minutes, possibly even twelve, considering the time it took to fall from a height of 5,000 meters. Over that period the passengers followed the command that is given by the crew in cases of even mild turbulence—to fasten their seat belts. Even if the corpses managed to completely rot away in their "watery grave," then, as investigators surmise, their skeletons should have remained, fixed to their seats.

Some will ask: why did the people disappear, while their things and the plane fragments remained on the bottom of the ocean? I will answer the question with another question: why did the body of Jesus disappear, while the shroud remained in the cave? Why are ships sometimes found in the Bermuda Triangle, all their seaworthy features intact, but without a single person? Apparently, because a mysterious hyper-field, depending on physical characteristics, selectively affects living and non-living material.

2.14. Was God an Astronaut?

One way or another, I encountered some information that is not subject to doubt. This event happened in 1933, when some strange maps of the world were discovered in a cartographic archive in Turkey. They had been compiled in 1513 by the Turkish Admiral Piri Reis, and turned out to be copies of Greek maps from the time of Alexander of Macedonia.

The outlines of America and Antarctica had been drawn on these maps! But somehow strangely: South America and the western coast of Antarctica are joined by an isthmus, while Antarctica itself is depicted with mountain ranges and coastal islands, interior reservoirs and rivers—in a word, as though this continent were not encased in a shell of ice.

Many millions of years ago, Antarctica was indeed ice-free, and was joined by isthmuses not only with South America, but also with Australia. That was the Miocene Epoch, when simian-like primates were just beginning to appear on Earth…

Open a biology textbook, esteemed reader, or even better, visit a museum of paleontology. Go to the exhibit on human evolution, and look closely at the bust of Pithecanthropus Erectus. Look at the expression on its face. Attentively, please…And now ask yourself the question: could the possessor of such a face produce even a rough plan of the terrain he inhabits? Of course, this is highly doubtful.

All the more so, it's impossible to believe that some ancestor of Pithecanthropus Erectus, even an extremely gifted one, could have put together a map of the world. One must concede that the strange details on ancient maps are simply the fruits of their creator's imagination.

In 1952, the first echolocation investigations of Antarctica were conducted. Then the true picture of its coastal outlines, hidden under a several-meters-thick covering of ice, was first revealed to humanity.

Five years later, it occurred to Fr. Daniel Linehan, geophysicist with the US Navy, to compare new maps of the Antarctic with the maps of Piri Reis: the coastal outlines of the continent on both maps conformed to each other exactly. Later research showed that the mountain ranges on the ancient maps were depicted "fabulously accurately!"

This discovery elicited a real stir in the scientific world—after all, it followed from this, that the images on the maps of the Turkish Admiral had been copied in the past, more than once, and the original itself was made many millions of years ago! As they say, it's unbelievable, but it's a fact. The main sensation, however, lay ahead: it was found that the maps of Piri Reis conformed completely with images of the Earth taken by photographic means from an American lunar rocket. Conclusion: the images of continents on the ancient maps were made by a photographic process from cosmic space!

It's especially curious that, working on his map of the world, Piri Reis made use not only of maps made by the ancient Greeks, but also a map made by Christopher Columbus in 1492, before his voyage to the American continent! Columbus, as we may remember, collected ancient maps and, judging by everything, among his collection there was one that charted unknown lands. It seems that Columbus was in possession of true evidence about the lands which he later discovered.

Thus it becomes understandable, from where the seafarer drew his conviction of the correctness of his conception, why for a duration of nearly 20 years he remained faithful to the idea that his contemporaries found so crazy that even children, seeing him out on the street, would pointedly twirl their fingers near their foreheads. Today one can only be amazed at the courage of Columbus, and the stubbornness with which he approached his goal—after all, in those long-ago times no one knew how huge the ocean really was, or whether it was possible to sail across it.

Indeed, it borders on a miracle that the light, practically deck-less little caravels could withstand the frequent and violent Atlantic storms on their way to America and back to Europe. And can't the spiritual fortitude and unbendable will that helped Columbus quell a mutiny on his ships, sparked by the superstitious sailors' witnessing of the maritime spectacle of a pillar of fire, also be called a miracle?

Evaluating the numerous difficulties and dangers that Columbus encountered on the way to realizing his plan, you involuntarily come to the thought that he couldn't have overcome them without the help of a Higher Power—so unlikely seems the successful outcome of his venture. And here, with frightening plausibility, fantasies arise of extraterrestrial gods who, from time to time, select among humans those individuals capable of carrying out, in a manner not otherwise possible, THEIR will, to set human history on a course needed by THEM.

Was God an astronaut? The Swiss writer, Erich von Däniken, once gave serious thought to this question. He answers the question in the affirmative—yes, God was an astronaut. Or, more likely: those whom ancient peoples took to be gods were actually visitors from other stellar worlds. If one were to be completely accurate, however, one should say that historians, ethnographers, and archaeologists first pondered this question, and their conjectures were generalized and developed by Däniken.

In his books, which gained worldwide acclaim, Däniken indulges in fantasies regarding how these contacts may have been realized: "The astronauts quickly learn the language of earthlings with the help of contemporary technology...Part of the population would be trained to search out and deliver fuel for the return of the ship to its planet; the chosen one from among the favorites would become a ruler of his own kind. Possibly, he would be given a radio transmitter, with whose aid he may at any time communicate with the gods..."

Here I'd like to mention that the anecdote about the Jewish wireless telegraph cited at the beginning of the book finds in Däniken an explanation based in natural science. The writer asserts in all seriousness that the prophet Moses received pertinent instructions from extraterrestrials by means of a radio receiver. Däniken finds confirmation of this in the Old Testament.

Here is the plan proposed by the Lord God: "And they shall make an ark of shit-tim wood: two cubits and a half shall be the length thereof, and a cubit and a half the breadth thereof, and a cubit and a half the height thereof. And thou shalt overlay it with pure gold, within and without shalt thou overlay it, and shalt make upon it a crown of gold round about.

"And thou shalt cast four rings of gold for it, and put them in the four corners thereof; and two rings shall be in the one side of it, and tow rings in the other side of it. And thou shalt make staves of shit-tim wood, and overlay them with gold. And thou shalt put the staves into the rings by the sides of the ark, that the ark may be borne with them. The staves shall be in the rings of the ark: they shall not be taken from it. And thou shalt put into the ark the testimony which I shall give thee."[26]

Moses did everything according to the given plan, so that on their journey God could communicate with his chosen prophet from the roof of the ark. "And when they came to Na-chon's threshing floor, Uz-zah put forth his hand to the ark of God, and took hold of it; for the oxen shook it. And the anger of the Lord was kindled against Uz-zah; and God smote him there for his error; and there he died by the ark of God."[27]

Däniken comments on this passage in the Bible: "Clearly, the Ark of the Covenant was electrically charged! The thing is, if all the directives of Moses are precisely carried out to reconstruct the Ark anew, we would have nothing less than an electrical capacitor with a charge of a few hundred volts. The gold wreath served as a charger for the capacitor, constituted from gold platelets. If one of the gold cherubim on the roof is a magnet, then you have before you a complete electrodynamic loudspeaker, or perhaps something like an intercom for communication between Moses and the crew of the spaceship..."

Such a liberal interpretation of the Bible aroused indignation among believers, while people who understand something about radio technology note that the level of Däniken's knowledge in that field leaves a lot to be desired, to put it mildly. But for the armies of ufologists, Däniken's fantasies proved a real boon, elevated almost to the status of evidence: UFOs are the product of extraterrestrial technology.

The *civilizants*, in their turn, make use of this "evidence" in their attempts to resolve the Bermudan issue. First, they cite mysterious events and disappearances, emphasizing that even competent commissions are incapable of explaining their causes. Then they circulate tales of UFOs—dates, names of eyewitnesses, retellings of their stories, comments of "authorities"—in such quantities that even skeptics finally begin to think there must be something to it. And here their goal is achieved: for someone who starts to view UFOs as extraterrestrial spaceships, all Bermudan riddles acquire a logical explanation at once.

---

[26] Exodus 25:10-16 (KJV).

[27] 2 Samuel 6: 6-7 (KJV).

However, the thoughtful and discerning reader won't find stories of visitors from the cosmos or a parallel universe to be convincing. Such a reader will quickly figure out that, as keys to the solution of the Bermudan issue, he is being fed tales that are themselves in need of proof. In the next chapter I will present a few such tales for my reader to judge.

2.15. Aboard an UFO

In 1983 French information sources presented readers with the following story. One day Frank Fontaine, Jean-Pierre Privo, and Solomon N'die were getting ready to go to the market. Frank was already behind the wheel of his auto, while the other friends finished loading items designated for sale. Suddenly a bright spot of light appeared in the morning sky, slowly and silently gliding to earth and leaving behind it a slanted, glowing trail.

"It's a falling plane!" Frank cried out to his friends. "I'll go look, come after me!" He pointed in the direction of a nearby power station, and took off. When Jean-Pierre and Solomon ran up to the power station they saw an amazing scene: the front part of the car was obscured by a shining spherical cloud.

Then the cloud lifted and disappeared in the sky; Frank was not in the car. His friends turned to the local police precinct for help. Understandably, their account was met with skepticism; however, the search for Frank that was undertaken by the police didn't lead to anything.

In 1989, the Russian newspaper *Sovetskaya Molodyozh [Soviet Youth]* published the following story.

"In the Kirovograd region, a 58-year-old chauffeur disappeared for five days. His relatives got worried, and a search was mounted. But the chauffeur turned up in the yard of his own home and, by his own account, had only been gone a couple of hours. According to his tale, two creatures enticed him into some sort of dome with a spherical body.

"Inside there were armchairs and windows. His native village began to fall away, and stars appeared all around. In this manner he ended up on another planet, where the climate was temperate, green trees with pink flowers were growing, and there were buildings similar to those on Earth, with shining crosses towering above them. There was a detailed dialogue, but he couldn't remember much. Then he was returned to Earth..."

The only people who could believe a similar story are naïve women whose "irresponsible" husbands disappear from home for a few days. True, Frank's friends didn't see extraterrestrials either. But they did see a shining spherical cloud hovering over his auto. Anyway, Frank, like the 58-year-old chauffeur, showed up himself. After a week.

He appeared on the porch of his friend Solomon's apartment and asked: "Why have you changed into your pajamas, when we're supposed to go to the market?" Later, Frank related that, when he got to the power station, his car's engine died, and he fell into a swoon.

"…I came to in a laboratory. It was a room with white walls, kind of like auditorium. There were lots of different kinds of machines. There were glowing dials everywhere…They began to talk to me. These are very smart, very wise people…"

If it's possible to guess at the reasons compelling a 58-year-old chauffeur to "disappear" for a few days, it's harder to imagine what could have made Frank run off from his friends right in the moments when he and they were getting ready to go out and make some money. One can't actually think that a young man would hide out somewhere for a whole week for the sole reason of getting his friends to believe in a made-up meeting with extraterrestrials.

I'll mention one more story. On a winter evening in 1958, Brazilian farmer Antonio Boas and his brother were working in a field on a tractor. Suddenly they saw a huge, round object, reminiscent of an auto wheel. This object emitted a light so bright that it hurt the eyes. It was a bright red color and illuminated a large area. Antonio suggested to his brother that they go have a look to see what it was. But his brother declined.

Antonio went alone. The round object turned out to be a UFO on three metallic "legs." Further, the 23-year-old farmer tells of how extraterrestrials overcame his resistance and dragged him into their flying apparatus. However, we'll give Antonio the opportunity to tell the compelling story himself.

"…At last they were able to shove me into a small, square room. The flickering light of the metal ceiling was reflected on the polished metal walls; the light emanated from a multitude of tetrahedral lamps arranged under the ceiling.

"I was set on the floor. The entrance door, together with the folding ladder, rose and slammed shut, completely blending into the wall. One of the five creatures let me know by gestures that I was to follow him. I obeyed, since I had no other choice. Together we went into a different, oval-shaped room that was bigger than the previous one…In the room there were lots of twirling stools that looked like the kind we have in our bars. That way, everyone sitting on a stool was able to turn in different directions. They kept firm hold of me the whole time, and it seemed they were talking about me. When I say 'talked,' that doesn't mean in the slightest that I heard anything resembling human sounds. I can't repeat them.

"Suddenly it seemed that they'd made a decision. All five began to undress me. I defended myself, shouted and cursed. They stopped for a second, looked at me as though wanting me to understand that they were courteous people. But that didn't stop them from stripping me naked.

"In the process, they didn't cause me any pain, and they didn't rip my clothes. The result was, I stood naked and scared to death, since I didn't know what else they wanted to do to me. One of them came up to me, holding something like a wet mop in his hand, and started rubbing my body with liquid…

"Suddenly…through a door, two more came in. In their hands were two fairly thick, red, rubber tubes, each of which was more than a meter long. One of these hoses was attached to a goblet-shaped glass receptacle. On the other end there was a nozzle that looked like a glass tube. They applied it to the skin of my chin, right here, where you can still see a dark spot left over from the scar.

"At first, I didn't feel any pain or itching. Then this spot started to burn and itch. I saw that the cup slowly filled halfway with my blood. Then they interrupted their work, removed one nozzle, replaced it with another, and drew blood from the other side of my chin. That left a similar dark spot. This time the cup filled to the brim. Then they left, the door closed behind them, and I was left alone…"

I'll interrupt Antonio's story to let the reader know some curious details, typical of such tales. According to publications, in the USA alone, thousands of people insist that they were abducted by extraterrestrials, and subjected to medical examinations on their spaceships. Analysis of these reports is conducted privately by the scholarly group, "Visit."

"Our clients, as a rule, are nervous people," says the group's psychologist. "Many have damaged hair, eyesight, digestion, there are traces of burns most likely resulting from radiation. The abducted…beg us to answer, could it really be that what happened to them is true? It's hard to accuse them of deception. We randomly checked up on a few people—they had never had dealings with surgeons. Yet the scars strongly resemble scars that result after surgery…"

In eyewitness accounts, details such as "pillar of light" are repeated. "It would flash, and the 'victim' would appear inside a round hall with scattered light, seemingly emanating from the sloping walls…People would lose their sense of time, experience weightlessness and total helplessness. Their abductors are humanoid. They exhibit neither violence, nor particular gentleness—they impose a trance and start conducting various biological examinations. Among these are also experiments in reproduction."

However, let's let Antonio Boas finish his story.

"…Quite a bit of time passed, no less than half an hour, but no one remembered about me. There was nothing in the room except a large bench without a headboard, covered with a thick, soft, gray material. Because I was pretty tired after all these anxieties, I sat down on this bench…

"I was extremely depressed. What else did fate have in store for me? Up to this point, I hadn't put together the slightest image of what these creatures really looked like. All five were wearing tightly fitting jumpsuits made of thick gray material that was very soft. On their heads, they had helmets of the same color. These helmets concealed everything except the eyes, which were covered with glass resembling spectacles. The sleeves of the jumpsuits were long and narrow.

"The wrists and hands, with five fingers, were hidden inside thick gloves of the same color, that doubtless restricted their movement, though this didn't prevent them from keeping a tight hold on me and manipulating the rubber hose that drew my blood. There were no buttons or pockets on the jumpsuits. The trousers were snug, and extended right into the shoes, which were like sneakers. In any case, they were dressed differently from us…

"After some time, which seemed like an eternity, a rustling at the door distracted me from my thoughts. I glanced around the room and saw a woman slowly approaching me. She was absolutely naked, just like me. I was speechless, and it seemed the woman was amused by the expression on my face. She was very beautiful, but with a beauty that was completely different from the women I was accustomed to meeting.

"Her hair was soft and light, even very light, as though it was bleached, parted in the middle and falling on her back in locks twisted inward. She had big, blue, almond-shaped eyes. Her nose was straight. Unusually high cheekbones gave her face a unique form. It was wider than that of South American Indian women. Her sharp chin made her face seem triangular.

"She had thin, slightly prominent lips, and her ears, which I looked closely at only later, were just like the ears of our women. Her body was strikingly beautiful: wide hips, long legs, a small foot, narrow wrists, and normal nails on the fingers. She was much smaller than I.

"The woman silently approached me and looked at me. Suddenly, she embraced me and started rubbing her face against mine. Alone with this woman, I was very aroused. This may seem improbable, but I think the reason for this was the liquid they had rubbed on me. They probably did it with an intent. In spite of this, I wouldn't trade a single one of our women for her, because I prefer women I can talk to, and who understand me. She was making these grunting noises that completely bewildered me. I was totally pissed.

"Then one of the ship's crew members came with my clothes, and I got dressed again. Except for my lighter, nothing was missing. It's possible I lost it during the scuffle. We returned to the other room, where crew members were sitting on the twirling stools and seemingly conversing. While they 'talked' to each other, I tried to accurately remember all the details of my surroundings. At that time, I noticed a square box with a glass lid that stood on a table. Under the glass there was a disc that looked like the dial of an alarm clock, but with black markings and only one hand.

"Then it occurred to me: I had to steal this object; it would be proof of my adventure. I started to move cautiously toward the box, taking advantage of the fact that they weren't looking at me. Then I quickly snatched it off the table with both hands. It was heavy, not less than two kilograms in weight. But I didn't get a chance to look at it closer: one of the sitters jumped up, pushed me aside, angrily tore the box out of my hands, and put it back in its place…It became clear to me that they were friendly toward me only when I behaved myself. So why risk it, if there was nothing I could do anyway…

"At last, one of the crew got up and let me understand that I was to follow him…We approached the open doorway with the ladder already lowered, but didn't go down. I was ordered to stand on the platform that stretched to either side of the doorway. It was narrow, but you could follow it all the way around the vehicle. We went forward, and I saw a square metallic protrusion extending from the vehicle; on the opposite side was another, exactly like it…

"Above the platform I could see countless tetrahedral lamps, mounted into the body of the vehicle. Their reddish light illuminated the platform, which ended in front of me with a large disc of thick glass. Apparently, this disc served as a porthole, though from the outside it seemed completely opaque. My guide pointed up, where a huge dish-like dome was rotating. As it slowly moved, it was constantly illuminated by a green light whose source I couldn't determine. A particular sound was connected with the rotation, reminiscent of a vacuum cleaner…

"I was finally led to the metal ladder and given to understand that I could go. Finding myself back on the ground, I looked up again. My companion still stood there; he pointed at himself, then at me, and then at the southern part of the sky. Then he made a motion for me to move to the side, and vanished into the vehicle.

"The metal ladder gathered itself up, the steps sliding one into the other; the door rose and moved into the wall of the vehicle. The glow of the projector and dome became brighter and brighter. The vehicle slowly rose, maintaining a vertical position. At the same time, the landing gear stowed itself away, and the lower part of the apparatus became completely smooth.

"The object continued to rise higher; at 30-50 meters off the ground it paused for a couple of seconds, during which its illumination increased, the buzzing became louder, and the cupola began to rotate with incredible speed. Lurching slightly to the side, with a rhythmic knocking sound, the vehicle suddenly tore off in a southerly direction, and in a few seconds disappeared from view. And then I returned to my tractor…"

2.16. Ufology, the Sister of Parapsychology

The stories I cited in the previous chapter were not selected randomly. As a matter of fact, they contain identical details. For example, this coincidence: to the 58-year-old chauffeur who made a trip to another planet, it seemed that he'd only been gone a couple of hours; Frank didn't even notice the gap in time.

A disruption in the sense of time is often mentioned by people who encounter UFOs. In the story of the Brazilian farmer, there are descriptions of details that are encountered in many stories of UFOs: protrusions (cones) on the bodies of the flying apparatus, the clothing of the extraterrestrials – silvery, skintight, minus buttons or visible seams.

Other details from eyewitness stories of UFOs are repeated so often that they can't help but draw the attention of ufologists. They consider that it's precisely these "correlations" that constitute proof of the veracity of such stories. However, belief in their authenticity is strongly undermined by the mass information media. Why, these stories are immediately taken up by a huge army of journalists and, like a flu pandemic, spread over countries and continents.

They're reported from television screens, interviews with eyewitnesses are published in newspapers and magazines; and then these stories are snatched up by people writing entertaining books about extraterrestrials. Ufology literature is always abundantly displayed on library shelves and in bookstores, and planet Earth is not lacking in people wishing to become celebrities by any means.

And today there's no simpler or quicker way to draw attention to one's persona, than by telling journalists the next breathtaking tale of a meeting with guests from the cosmos. And to make the story even more convincing, it can always be seasoned with details appropriated from the narratives of other, similar "eyewitnesses." The temptation is even harder to resist because the risk of being exposed in deception is nil. But still, better to acquire a "witness" in the person of a close relative, or a wife (husband). And if the scenario is well rehearsed, there's a big chance that the newly concocted story will become the domain of publicity, and its hero will become famous…

Returning home to Portsmouth on September 19, 1961 the American couple, Betty and Barney Hill, noticed that their auto was being followed by some kind of shining object. Betty took a look through binoculars. What she saw defies description: this was a huge UFO with a rotating dome and blinking lights.

"After the object began to hover in the air, they stopped their car…Suddenly the couple saw that the object was diving at their auto…And while it hung over them, they heard a series of short, loud, buzzing noises…" The UFO landed not far from the car, and the couple could see a few humanoid creatures through its transparent dome. Barney got out and wanted to go closer to the object, but Betty yelled, "Come back! They'll haul us away!"

Betty's apprehension was justified. As in the incident with Antonio Boas, the extraterrestrials dragged the people into their ship. At first the couple couldn't remember anything that happened to them on board the UFO. When they regained consciousness, they saw with surprise that they were driving in their car, 35 miles from the place where they'd met up with the cosmic ship…

The reader has probably noticed that in the reports about encounters with UFOs, I present different parts in the form of citations. The frequency of their interchangeability may give the impression that the author was too lazy to retell the story in his own words. But it's not like that. The thing is, precisely in these citations there are indications of details that, to my view, need further discussion.

Retelling similar stories carries the well-known risk, and even temptation, to specifically highlight and present in a favorable light those moments that correspond to the author's vision. That risk, of course, I'd like to avoid.

"…After the incident, both eyewitnesses began having nightmares that sent them, three years later, to a psychiatrist, Dr. Benjamin Simon, who put them under a regressive hypnosis to find out why the problem had appeared; only then was it discovered that the origin of the nightmares was connected with the loss of two hours. Under hypnosis, Betty and Barney separately reported that they had been taken by unknown creatures aboard a UFO…"

The word "separately" in the last sentence of the previous paragraph is obviously meant as evidence that the content of the story told by each spouse fully corresponded to the other, and that therefore both stories are true. Here, however, a qualification must be made: only if, between them, Betty and Barney hadn't made a previous agreement.

I do not, by any means, wish to expose the Hill couple in a scam. But, like an independent investigator, I'm attempting to grope my way to a different possibility by which evidence given separately by two people might correspond. And if such a possibility is discovered, then—in line with Occam's principle—the story of an unlikely event must automatically be subjected to doubt.

The fact that that the UFO was picked up on radar at a nearby military base also doesn't prove the veracity of the Hills' story. For an unknown object in the sky (ball lightning or lights on an airplane) could have inflamed the imagination of one or the other of the spouses.

Let's interrupt the discussion of UFOs for a time, and turn to the field of medical psychology. It may happen that, having lived through a psychic trauma, a person forgets about it, or in the parlance of psychoanalysis, displaces the memory of the trauma to the subconscious. Over time, he may develop a neurotic condition.

With the aid of hypnosis, it's sometimes possible to reconstruct the traumatic situation in the memory and, simultaneously implementing other healing methods, to rid the patient of the affliction. The adjective "regressive" is fully appropriate here, since it indicates a reference to the past.

Some psychoanalysts, convinced that the causes of neuroses were rooted exclusively in the early childhood of their patients, began to use regressive psychosis on them. Adults, immersed in a hypnotic state, simultaneously regressed into childhood: they threw tantrums, blew bubbles, asked to go on the potty. All this was regarded as signs of beginning recovery.

In the circles of parapsychologists taken with Freudianism, an idea once caught on, that nervous ailments are caused by psychic traumas experienced in a previous life, that is, when a person's soul inhabited another bodily shell. This idea was warmly met by some psychoanalysts, who decided to test it with experiments. With the aid of regressive hypnosis, they proposed to kill two birds with one stone: achieve the healing effect, and confirm the hypothesis of the migration of souls.

Subjects for such experiments are selected among people easily influenced and psychically unbalanced, among whom are often found personalities inclined to use any means by which to be the center of attention. These people are particularly flattered that they have been chosen by scientists, and that they have an opportunity to participate in research whose results may liberate millions from suffering.

…And so, the subject sinks into a soft armchair and closes his eyes. He is informed about the goals of the experiment and will naturally strive not to disappoint the expectations of the researchers. He allows himself to be brought into a drowsy state, so conducive to involuntary streams of consciousness. At the same time, the subject may deliberately play along with the hypnotizer, making up the history of his previous life during the course of the experiment.

Of course, plain old deception has its place here. But more often, the subject "dislodges" all doubts concerning the purity of the experiment. Within a few minutes, he is already "reincarnated": he presents a different character, living in the century before last, recounting everything helpfully supplied by his seriously overwrought imagination. Thus it happens, that experimenters and subjects participate in a strange game in which, impatiently awaiting positive results, they seem to fool themselves.

Try, however, to explain all this to enthusiasts of the idea of reincarnation, and in the spirit of counterargument they will cite publications in which it's reported how one little girl began speaking in Sumerian under hypnosis, and another little boy spoke Sanskrit. How similar publications affect the mind is easy to test. Tell one of your friends a fantastical story. More likely than not, he won't believe it. But if your friend encounters a written version of the same story in a thick journal, his attitude toward it will be much less skeptical. Such is the strength of the printed word.

A qualification: I have no basis for rejecting (and even admit) the possibility of the existence of the soul and its migration into various bodies. But what I've read about experiments that elevate this possibility to a proven fact, I categorically reject. I would like to, myself, select at random a dozen people, and bring them to the initiators of the idea of reincarnation. And if even one of the subjects begins to speak in a language unfamiliar to him, or at least correctly identifies the location of a secret hoard stashed somewhere in Central Asia by Alexander of Macedonia, then I'll surrender.

Ufology is the sister of parapsychology, and for this reason representatives of both fields adopt similar methods of justifying their theories. This is simply explained by the absence of other means of proof.

On October 12, 1989, the newspaper *Komsomol'skaya Pravda* published a chilling account by the journalist Pavel Mukhortov, about his encounter with representatives of an extraterrestrial civilization. "…During the day I sensed some sort of inner jolts—I was pulled in the direction where I'd seen the 'saucer.' At night I made up my mind to go.

The feeling was unpleasant, as though they're putting pressure on your legs, not letting you go…About 30 meters away I saw two glowing eyes…Then figures of two to four meters in height began to appear in even rows through the darkness. The insides of the figures were all voluminous and somehow illuminated. I began to mentally ask them questions:

- Where are you from?
- In the constellation of Libra, the Red Star—that is our home.
- What is your objective?
- That depends on the Center. We're controlled by a central system.
- Are you able to take me to your planet?
- That's…dangerous for us.
- What's dangerous about it?
- Bacteria of ideas…"

With regard to the conversations conducted with representatives of the Libra constellation, Emil Bachurin, a recognized Russian authority on communication with the extraterrestrial mind, announced: "I don't dare claim that a close encounter occurred. But there was a telepathic contact…"

In my youth, I got fascinated by parapsychology. I retain some interest in it to this day, though I know that not all information related to that field can be taken on faith. It's the same as the issue with UFOs: often, parapsychological "effects" are the pranks of jokers, the results of fakery or self-delusion.

It's impossible, of course, to disprove Bachurin's assertion that telepathic contact had taken place with an extraterrestrial mind. But no one is going to stop us from proposing a possible mechanism of self-deception. This mechanism may be tested immediately with the help of a simple psychological experiment.

Put aside my book, esteemed reader. Close your eyes, and mobilize your imagination. Imagine an extraterrestrial, specially come to Earth from that same constellation of Libra for a visit with you. He's three meters tall. He has burning eyes, and wears a shining jumpsuit, transparent to the point of immodesty. With some effort, he squeezes himself into the armchair you proffer, and crosses his legs. He refuses a cigarette, he doesn't smoke.

Have you imagined all this? Excellent. And now, mentally address some question to your guest. In less than a minute, you'll be convinced that the extraterrestrial's "answer" has floated into your head. In this manner, if the conversation appears to be substantial, you may carry it on as long as you wish...

Let's return to the story of the Hill couple. "...Betty continued looking through the book given her by the 'Captain.' This was a large book, filled with vertical lines—thin ones, thick ones, broken ones, crooked ones. The 'Captain' promised to give it to the woman to serve as proof of the reality of this unprecedented contact. Betty gathered her courage and asked the 'Captain' where they had come from. He ascertained that she knew about the Universe, but upon hearing that her conception of it was confined to what she'd learned from a high school course, he shook his head in disappointment.

Nevertheless, he got out a large album and held it out to his prisoner. The pages of the album, with pictures of a star-studded sky, had been manufactured by an unfamiliar process: they displayed a three-dimensional, holographic picture! 'Like you were looking through a window,' Betty recounted later.

One map intrigued her particularly; on it were lines connecting the stars. The 'Captain' explained that the thick lines designated trade routes, the thin lines were occasional expeditions, and the broken lines indicated future flights.

'Where is your star?' Betty asked.

'Do you know where your Sun is on this map?' came question in response.

'No, I don't know.'

'Then how can I explain to you where we're from?'"

Ah, the star map contains trade routes! This means the cosmos is inhabited by rational creatures! This means that contact between worlds has long been established! But why aren't we Earthlings drawn into the cosmic association of civilizations? Or is it that we, the children of Earth, are so "undeveloped" that the extraterrestrials are withholding information about themselves, the way some parents withhold from their kid the mystery of childbirth?

But now the captain of the cosmic spaceship gives Betty the star catalog. Is this not a sign that now humanity is entering a level of maturity, and that it's finally time to get acquainted? But why did the extraterrestrials pick for this the first people they happened to encounter on the road? Wouldn't it be more intelligent to address the United Nations, or the government of one of the superpowers? Yes, there's little logic in all of this. Let's see, however, how the event developed further.

"…Prior to releasing the prisoners, the 'Captain' announced that the crew objected to giving them physical evidence of the contact.

'You have to forget everything that transpired,' he said, turning to Betty.

'No, no. I'll never forget it, never!' the woman contradicted him petulantly.

'You'll come back again?'

'That does not depend on us. Possibly, we'll meet again.'

'But how will you find us among millions of people?'

'We always find the ones we need,' the 'Captain' replied pointedly…"

No matter how far removed from human logic extraterrestrial logic may be, even it must conform to certain universal laws. Is it possible to believe that a matter of such paramount importance as contact with another civilization is conducted on a whim, that every step of such an enterprise is not discussed and enshrined in special instructions that all members of a stellar expedition are bound to observe without deviation? Why did the captain of a cosmic spaceship encounter differences with his crew in this regard?

If I were to find myself aboard a UFO, I'd definitely want to snatch some physical evidence of the fact. For this reason, I find it easy to imagine the vexation that must have seized Betty when she was forced to return such a fancy item as the album with maps of the starry sky. Heck, the woman didn't have enough resolve, like Antonio Boas, to swipe something from the extraterrestrials. Still, as we remember, when Antonio attempted to steal some object off a table, he was seized by the hand.

It's strange, of course, that among hundreds of people who have been aboard a UFO, or even visited other stellar worlds, not one has been found who managed to present to humanity even one little extra-Earthly object. But ufologists extract evidence of the veracity of these stories with the aid of regressive hypnosis.

"…During the course of one of the hypnotism sessions at Benjamin Simon's clinic, the researchers got a simultaneous idea to try to pull from

Betty's long-term memory the star map she was shown aboard the UFO. The practice of regressive hypnosis has made similar discoveries. Dr. Benjamin Simon enthusiastically began to experiment…"

The "experimenters" were, of course, ufologists, or more accurately, members of the ufological association, National Investigations Committee on Aerial Phenomena (NICAP), headed by retired Major Donald Keyhoe, among others. It was he who had introduced the couple, Betty and Barney Hill, to Dr. Simon.

"And so?" you ask with hidden hope. "Did the doctor succeed in drawing from Betty's memory at least a general outline of the positions of stars on the map?"

What general outline?! "The success [of the experiment] exceeded expectations. Over the course of several sessions, Betty Hill drew a map; moreover, each time she added some details or changes to her chart. When the work was finished, the unique map was published in newspapers for the knowledgeable opinion of specialists.

"From the completed map it followed that only from the Net Constellation would it be possible to see a similar placement of stars. The distance to the visitors' base was colossal, over thirty light years! How could they manage to get here? There can only be one answer: they have the ability to travel at near light speeds…"

And so, having determined where the extraterrestrials had come from, the ufologists put a period on the story of the Hills. But I still have questions: by what criteria did the researchers determine that it was the published version of Betty's map that was the "completed" one? And further. Why seek specialists through the newspapers? After all, for a knowledgeable opinion it would have been possible to turn to any astronomical association, or to an observatory.

The "knowledgeable opinion" was actually expressed by Margery Fish, a school teacher from Ohio. Believing the story of the Hill couple, the teacher determined to find the "homeland" of the visitors on a stellar map. And – she found it!

Let us try to draw a map of the stars in the sky. On a clean sheet of paper, we'll freely place a few dozen dots. And now we'll consider the question: does the position of the dots on the sheet of paper correspond to the view of the starry sky that opens up to the eye of an observer on the surface of some planet? "But there are billions of stars!" the reader will exclaim here. "And many more planets!" Well, this means that our imagined observer would have to make journeys in turn to all the stars visible through a telescope, and make photographs of the sky from both poles of their planets. The resulting album would be imposing, to be sure – hundreds of millions of maps! But now it will be possible to compare these maps with the picture we drew. And, as pointedly noted by one author, "With billions of stars in the expanses of the Milky Way, any drawing with any configuration of stars will always find correspondence." In seeking this "correspondence," the school teacher from Ohio spent only five years.

2.17. History Repeats Itself

Ufologists have estimated that approximately 5 million people from different countries encountered UFO phenomena over the course of the 20th century; "if we suppose that half of these 5 million were hallucinating, that still leaves over a couple million." That is, they attempt to convince us that not all people speaking of encounters with extraterrestrial spaceships and their passengers are jokers, crooks, or mentally ill. However, for me, even two and a half million witnesses don't make an argument. Once upon a time, the courts of the Inquisition received numerous testimonies regarding contacts with the devil, and the sorcery of witches. Now it's certain: this was mass hysteria, gripping the population of medieval Europe. History, as they say, repeats itself…

Back in the 19th century, the Russian bishop Ignaty (Bryanchaninov) remarked: "We're gradually coming to a time when a vast arena will open up for numerous and amazing false marvels. The marvels of the Antichrist will for the most part take place in the realm of the air…" These words turned out to be prophetic. Our contemporary, the French astronomer Jaques Balle, writes: "Observations of unusual events have suddenly washed over us by the thousands…A powerful force that influenced humanity in the past is once again affecting it in our day."

One of the victims of this "powerful force" was the mother of one of the pilots of Flight 19; out of her mind with grief, the woman convinced herself that her son had been abducted by aliens, that he was alive and living on another planet. The propagation of similar fancies is facilitated by ufological literature. Proofs? If you please. Here are the results of sociological surveys: by mid-20th century, 28% of the citizens of West Germany believed that our planet is visited by rational creatures from other worlds; a quarter of a century later, that number increased to 40%. Over 50 million Americans are convinced that extraterrestrials frequent their continent.

The fact that human nature is perhaps more puzzling than psychologists surmise, is attested to by the following fact. At one point, the Florida Insurance company offered to insure Americans against possible abduction by cosmic visitors. In a short time, it had 1,200 takers.

And so, childish fantasies of witches flying on broomsticks have been increasingly supplanted by visions of extraterrestrial spaceships. Yes, humanity has matured…

One of the main difficulties faced by investigators of UFO phenomena is the selection of criteria by which to classify the information submitted to them. Let's say that someone saw a glowing object in the sky. Is it a spaceship? Ball lightning? A meteorological probe? How can one know? One can't. However, if the eyewitness managed to get a photograph, then researchers have an easier time figuring out whether this was an illusory perception of a natural phenomenon, a manmade terrestrial object, of a "real UFO."

One such photograph – a slightly elongated semi-transparent disc – was published in 1979 by the magazine *Nauka i zhizn [Science and Life]*. After experts had judged the photo to be authentic, its "author" sent a letter to the editors, confessing to a hoax: "…I rolled the soft part of some black bread into a little ball, and stuffed it into the round hole of the sort of glass rosette you stick dinner candles in…Then I poked a match into the soft bread ball, tied some transparent fishing line to it, and hung this contraption from a balcony ledge against a background of early evening clouds."

The *Blue Book*, a rather well-known American collection of UFO sightings, contains many photographs of "flying saucers." "The majority of photos in the *Blue Book* archives are straightforward falsifications or manipulations. The images on some of them resemble hats, upside-down cups, steaming saucers and plates…and a whole arsenal of other familiar or lesser-known items. In many cases these are photos made in the spirit of a spoof or joke. There are also other kinds of photos – of actual phenomena or completely authentic objects. Such are photos of planes, balloons, meteors, stars, reflections of the moon, street lights, auto headlights and other sources of light; all of these, at the moment they were snapped or after development of the film, took on an unusual appearance, and so were taken for UFOs."

The quotation above by no means belongs to an opponent of the idea of cosmic contacts. It was made by the aforementioned American ufologist J.

Allen Hynek, who once declared, "I wouldn't have touched the subject of UFOs for anything, if I didn't consider that UFOs truly exist."

Thus, the majority of UFO images in the *Blue Book* are fakes. But what is represented by the minority? It's a reasonable question, given that the photograph published in *Science and Life* was considered authentic by experts…

On April 1, 1950, the German newspaper *Wiesbadener Tagblatt* ran a photo of a UFO passenger who had suffered a wreck, as readers were informed, near the town of Wiesbaden. The photo showed a humanoid, waist high to the two American officers accompanying it. The extraterrestrial wore an apparatus resembling a gas mask on its face. Only a few years later did readers find out that this April Fool joke had been concocted by the newspaper staff; the role of the "extraterrestrial" had been played by a five-year-old girl.

In 1955, an announcement proliferated in ufological literature, attributed to Douglas MacArthur, former commander of joint US forces in the Pacific. "The nations of the world must unite," the announcement said, "because the next war will be an interplanetary one. The Earth's nations will be obliged to form a unified front against invaders from other planets."

The authors quoting these lines cited the New York *Times* of October 9, 1955, but for a long time no one thought to look at the paper. It was later determined that these lines were not to be found in the entire New York *Times* file for 1955.

70 years ago, a report flew around the world concerning corpses of extraterrestrials kept in a secret hangar in the US, near the site of a "flying saucer" crash. They say that the body of one of the deceased humanoids was even shown to reporters. But later it became known that this was actually the mummy of a monkey that had earlier been shaved and painted blue for the sake of the practical joke.

In 1966, at the instigation of the US Air Force, a commission was created at the University of Colorado for the purpose of synthesizing an array of UFO sightings, and rendering an opinion. The group of 37 specialists in various fields was headed by the major physicist, Professor Edward Condon. The commission came to a discomforting conclusion: "A thorough study of the materials in our possession has allowed us to conclude that further investigations of UFOs are unlikely to be justified by the hope that they will contribute anything useful to science."

Ufologists obviously don't like such conclusions. They consider that the absence of proof cannot be "proof of absence," and warn against hasty decisions to wrap up the work of studying these phenomena. They say that the facts of the unmasking of false testimonies regarding extraterrestrials don't in themselves preclude the possibility of their presence on Earth, that is, that along with fabricated stories there may also be genuine ones. Another argument: the topic of UFOs is curated by military departments that classify everything. At the same time, special units of the defense services engage in disinformation – spreading laughable reports about contact with extraterrestrials, with the goal of creating a skeptical attitude toward the subject among the public. They write, for example, that the CIA promoted the publication of books written by critics of the hypothesis that extraterrestrials exist.

All of this may be so. But, on the other hand, to what point does one conduct searches? Where is the line that, have overstepped it, one can firmly say: "That's it, the investigations are complete, the result is negative!" Such a line is impossible to determine. Thus it happens, that in the eyes of world society ufologists constantly appear in a favorable light. On the one hand, they come out in favor of continued investigations, which can't but elicit sympathy; while on the other – all official pronouncements denying the presence on Earth of extraterrestrials they characterize as the unwillingness of the powers that be to sow panic among the people. Under the circumstances of an unending torrent of more and more new sightings of UFOs, all discussions about whether they appear as space ships or something else may be carried on indefinitely…

Part III
The Voice of Silence

"…Why is it I have these constant migraines? I think, maybe from the gunfire in France? You know, there's a vibration in the air. Of course, I can't hear gunfire, which is understandable. But it occurred to me that vibrations of the air could cause these effects."

A.J. Cronin, *The Stars Look Down*

3.1. Did the Pilots of Flight 19 Develop Mental Disorders During the Flight?

The region of the North Atlantic that encompasses the Sargasso Sea is also called the "Horse Latitudes." This name, Kusche explains, "arose back in the times when vessels with horses on board were becalmed here for long periods. Days would go by without even a hint of rain or wind, and supplies of potable water would dwindle catastrophically. Horses, crazed with thirst, would often break their tethers and throw themselves into the water. And it also happened" – concludes Kusche – "that the people themselves threw the weakened horses overboard, to save what was left of their water for the stronger and sturdier."

Kushce's attempt to explain the origin of the name "Horse Latitudes" is apparently predicated upon having readers who will take anything they read on faith. Personally, I find it hard to believe, in those long-ago times when there weren't refrigerators or canned goods, and sea voyages lasted for weeks, that when not only water but food supplies were running out, they would toss overboard something so precious as fresh horse meat. And anyway could normal people not give water to an animal so long that it, "dazed" with thirst, jumped overboard? Wouldn't it be more humane to finish it off in time for the benefit of both parties concerned?

However, another explanation for the "suicide" of horses can be found in the literature on the Triangle. It is associated with a long-standing belief among sailors that some mysterious force in these waters "has sporadic effects on the psyche".

"No tale of the ocean's mysteries would be complete without the story of the *Mary Celeste*. Although she was found, minus people, in the ocean between the Azores and Portugal, she is remembered most often in connection with the mysteries of the Bermuda Triangle. All vessels abandoned by their crew, wherever they may be found, are compared with the *Mary Celeste*, and all mysterious tales, whoever they happened to, invoke her name…"

So begins the unmasking of another Legend by Lawrence Kusche. The abandoned brigantine was discovered in the ocean on December 4, 1872. A conjecture was expressed that spoiled food had caused hallucinations among the crew, and people began throwing themselves into the ocean to escape the horrible apparitions. But Kusche offers his favorite explanation: "the vessel encountered a storm, and when it seemed to the crew members that they were about to go under, they lowered the lifeboats, and apparently all perished."

The *Mary Celeste* was found outside the Bermuda Triangle. But if Kusche had, as in the course of other incidents, charted on a map the route of the vessel (New York to Genoa), then it would have been clear that it had crossed the Bermuda Triangle.

In a like manner, regarding Taylor's reluctance there later emerged all sorts of conjectures: the lieutenant didn't explain why he didn't want to participate in the training flight, and to many this seemed strange. (Could he have had a premonition?) Among others, there was the speculation that Taylor possibly didn't feel well, or wasn't sober.

If the leader of Flight 19 was experiencing some sort of ailment, he had no reason to hide it. Did Taylor show up for work that day in a drunken state? Definitely—no. As indicated by Lieutenant Curtis' testimony, there wasn't anything strange about Taylor's behavior, he conducted himself normally in all respects.

The board of inquiry had reason to believe that during the training flight (or a few hours before it?) the members of the 19th link had mental disorders. Kusche, however, is of a different opinion.

The former pilot believes that "the radio messages which, in spite of interference, could be picked up on the ground, although they indicated the plight of the 19th link, did not indicate any mental disturbance in the crew members".

"Was this actually so? Let's recall: at the request of Lieutenant Cox, the commander of Flight 19 gave the call signals of the plane which he had flown previously in Miami. We can say, of course, that Taylor simply misspoke. But let's look further. To Lieutenant Cox's suggestion that he engage his emergency locator system, he replies that his plane lacks such an apparatus. Perhaps this was Taylor's first time flying a TBM-3 Avenger? No, out of 2,500 hours flown by him, 600 were on planes of this type, equipped with JFF devices.

Truly, isn't it strange: the commander of Flight 19 not only forgets to engage the apparatus, but announces that there is no such apparatus on his plane! When Lieutenant Cox attempted to reorient Taylor via radio, he himself was overflying the airbase on the coast of Florida, where the weather was "clear" and the sun shone. The weather began to deteriorate later—a huge, dark cloud was advancing. toward the region of the incident. Since the cloud was moving from the west, it had to first block the sun over the head of Lieutenant Cox. This means that, in the field of visibility of Flight 19's pilots, the sun was shining!

Let's imagine a school boy, deep in the forest. The compass that he prudently took with him has been accidentally broken. He has no wrist watch. The boy knows with certainty that his home is located west of the forest. But how to determine, which way is west? Fortunately, it's still light, and rays of sunlight penetrate the forest canopy. The sun—there's the guide! And if the boy has (even an approximate) idea of the time of day, then he can determine the direction of west in a matter of seconds.

Is it possible that the military pilots didn't know what is known to every school boy? That is incredible. All the same, the pilots of Flight 19 were unable to orient themselves by the sun. Judging by everything, along with the loss of their spatial orientation, their perception of time was disrupted as well—a clear indication of psychic impairment.[28]

I get uncomfortable chills up and down my spine every time I try to imagine what was going on in the Bermuda Triangle on December 5, 1945. Fourteen courageous pilots of the US Navy, instead of flying west to lifesaving shores, depart toward the open ocean on their last drops of fuel…"

---

[28] Grey Walter, an English physiologist, cites a curious example of how, in circumstances of desynchronization, humans lose their perception of time: "One subject said that he had been 'displaced in time to the side--'yesterday' appeared on the right, instead of being behind, while 'tomorrow' seemed to be to the left." Here the subject describes the disruption of the perception of time in terms of spatial categories. And this is not just metaphor. Time is one of the coordinates of physical space, in reality its integral component. Notably, biological clocks and certain forms of orientation share common mechanisms.

3.2. Robert Wood's Inaudible Note

I noted above that in my youth I was fond of parapsychology. Those were the sixties of the twentieth century, when the first swallows of freethinking began to appear in the domestic press – articles about amazing experiments carried out in the USSR and Western countries on the transmission of thoughts over a distance. Finding publications in magazines and newspapers on a topic of interest to me, I cut them out and folded to a special folder. I was impatiently waiting for a message (it must happen sometime!) about the nature of that mysterious factor that "transports" thought directly from brain to brain.

Over time, the folder became very swollen, I had to start a second one. On the cover, I printed in large letters—"EXTRA-SENSIBLE PERCEPTIONS." The folder also began to be replenished with information about the premonition of natural elements by people and animals, about "horse latitudes" and ghost ships aimlessly plowing the ocean expanses. So the topic of extrasensory perceptions intersected with the mysteries of the Bermuda Triangle…"Each one of us, apparently, harbors a secret wish of preserving a reverential thrill before phenomena that seemingly don't yield to logical scientific explanation." So writes Lawrence Kusche. It is he who attempts to kill off in us, in his analysis of 53 events, this essential for humanity aspiration to commune with a Mystery. Incidentally, why 53?

Charles Berlitz, in his book on the Bermuda Triangle, quotes a roster of 143 mysterious events. And David Grope quotes even more—211. Without a doubt, if one were to dig around among the materials gathered by the destroyer of mysteries, it would be possible to find many events that do not, seemingly, yield to "logical scientific explanation." These events, of course, didn't enter into his book…

From the report of the investigative commission:
Facts 8 – 12.

" In each plane except Gerber's there was a crew of three, including the pilot. In Gerber's plane the crew was one person short."

Why was the crew one man short on Gerber's plane? Perhaps that man was sick? But it's possible to suggest something else, for example that this crew member, pleading illness, in reality possessed information about preparations of sabotage. However, why guess—the security service probably had a talk with him. It turns out that this member of Gerber's crew, Allen Kosnar by name, "requested to be relieved of the flight for reasons of intuition." And his request was honored.

Excuse me, but why is Lawrence Kusche silent about this fact in his book? Isn't that to keep the reader away from another mystery: a premonition of disaster arose in two people simultaneously. And the disaster took place…

Recall also the story of the schooner Carroll A. Deering, captained by 66-year-old Wormwell. The ship made her way to South America and back to Diamond Shoals, but did not arrive at her destination port. Six months later, the sailing ship was found in the ocean without damage, but without a single crew member aboard.

Here is a remarkable detail in the event, as cited by David Grope in his book: "A few days before returning home, Warmwell wrote a letter home to Norfolk in which he shared with his loved ones the anxious thoughts that haunted him—as if his life hung in the balance and was about to be cut short."

Is this where intuition comes in? But what lies beneath this ill-defined word? Clairvoyance? Or unconscious perception of some physical factor, causing a feeling of anxiety, and sometimes panic attack, bordering on psychosis?

Youth is presumptuous, and I seriously decided to solve these mysteries. Baku Akhundov Central Library became, for me, one might say, a second home. In those days, when my fellow students, preparing for exams, were poring over anatomy and biology textbooks, I spent hours taking out chapters from books on oceanography, meteorology, psychophysiology…

In 1934, the Russian psychiatrist, M. Nikitin, described episodes of epileptic seizures triggered by the sounds of an organ. Apparently, the seizures were caused by infrasound—after all, along with the tones of music perceived by the ear, there always appear low, inaudible vibrations of the air caused by the vibration of organ pipes[29]. One can say, of course, that a purely emotional factor operates on the psyche—the music itself. However, a story connected with the name of the American physicist, Robert Wood, is evidence that the matter doesn't always lie in the music…

At the old Lyric Theater of London, a play was being staged in which the time of the action changed in one act from contemporary to the long-ago past. Producer Gilbert Miller tried in vain to achieve the necessary mysteriousness—all the tricks of the costumers and decorators seemed like pathetic gimmicks. Among some friends present at the theater during rehearsal was the physicist Robert Wood, who suggested creating an impact on the audience with a low note.

His idea was that a very low note, almost inaudible but vibrating the eardrum, would create in the spectator a mood of anxiety and a sense of mystery. A "superpipe" was constructed according to plan, and affixed to an organ located behind the scenes. They decided to test the "superpipe" during the next rehearsal. Only Miller, Wood, and a few other people in the house knew about the impending experiment.

The physicist apparently made a mistake in calculating the parameters of the pipe—there was absolutely no sound. But the effect was similar to that preceding an earthquake. The crystal in the chandeliers of the old Lyric rang out, and all the windows rattled. Panic ensued not only inside the theater, but also on adjoining Shaftesbury Avenue.

Employees of the electroacoustics laboratory at the Maritime Research Center in Marseille began complaining of headaches, nausea, and unmotivated anxiety. Several people went to doctors, but no pathology was found in them. One day, Professor V. Gavro, the head of the laboratory, put his palm to the wall and felt it shake slightly. It turned out that a small industrial enterprise, in which the exhaust pipe of the ventilation system generated low-frequency sound vibrations, started working near the center of scientific research.

It was here that the professor suspected that inaudible infrasound might be responsible for the poor well-being of his colleagues. It was decided to test the assumption with an infrasonic "whistle" operating on compressed air, similar

---

[29] A person hears sounds that lie in the frequency range – from 16 to 20 thousand vibrations per second. Those sound vibrations that are below the threshold of perception of the human ear (up to 16 hertz) are referred to as infrasounds, above 20,000 hertz – to ultrasounds.

to an organ pipe. "The first experience almost cost us our lives," Gavreau later recalled. Fortunately, we immediately turned off the generator. For several more hours we felt completely shattered. Everything was vibrating inside us—stomach, heart, lungs…In the neighboring laboratories, people were screaming."

No wonder they were screaming. As further research has shown, our internal organs, consisting of cavities, have their own frequencies of vibration, in the range of 2-15 hertz. Exposure to the same sound frequencies causes resonance of the stomach, heart, lungs, from which these organs begin to vibrate, which is accompanied by strong pain sensations. And the coincidence of the ranges of infrasonic frequencies with the rhythms of the brain biopotentials causes mental disorders. Thus, when exposed to infrasonic frequency of 5-7 hertz, volunteers experience unmotivated anxiety, fear, panic, "up to insanity".

The dissemination of such information by the media gave rise to the "voice of the sea" hypothesis. The name is connected to a discovery made by Soviet meteorologists in 1923. The icebreaker *Taimyr* was conducting a research voyage in the Arctic. Wind speed and direction at different heights were studied by sending meteorological balloon probes filled with oxygen into the atmosphere.

Once, in clear weather, one of the aerologists accidentally touched an inflated balloon probe to his face and felt sharp pain in his ears. A strong storm erupted that night. The then-young oceanographer V. Shuleikin, who was on board the ship, determined that a few hours before the onset of the storm, the skin of the balloon probe began to vibrate at a frequency of 6 – 13 Hertz.

In this manner it was proven that infrasonic oscillations, arising in the regions of storms from the friction of air against the crests of waves, resonate inside the balloon probes; at the same time, frequencies of 6 – 7 Hertz dominate the sound spectrum. The infrasound generated by the storm was called the "voice of the sea."[30]

---

[30] I'll cite a dramatic example of a maritime event that demonstrates how rescue operations are conducted in stormy weather. On October 9, 1913, a disaster took place in the Atlantic on the English steamship *Volturno*, with 564 passengers aboard. A fire broke out in the hold, and quickly engulfed the whole ship. Several vessels responded to the SOS signal, among them the Russian ship *Tsar*. The journal *Niva [Cornfield]*

The hypothesis of an insidious killer of sailors seemed quite plausible. The body of the ship becomes a secondary source of infrasound if its resonant frequency coincides with the frequency of infrasonic oscillations emanating from the region of an oncoming storm. Since sound travels through air at a speed of 1,200 km/hour, and through water at 6,000 km/hour, it reaches vessels faster than the storm. Supporters of this hypothesis call attention to the circumstance that many tragedies have occurred in clear weather, at the same time as a storm was raging, possibly hundreds of kilometers away from the place of occurrence.

One of the authors paints the following picture: "A quiet, calm ocean. Nothing portends of danger. But somewhere a storm is raging. And now, subject to the 'voice of the sea' born of this storm, the masts and bulkheads unexpectedly begin to vibrate and break apart, the sails are torn. The sailors, gripped by a sudden terror and brutal pain, have only one thought—get the hell off this cursed vessel! A little time goes by, and over an ocean as placid as before, floats a new *Flying Dutchman.*"

From an interview by a correspondent of the newspaper *Sotsialisticheskaya Industria [Socialist Industry]* with Professor V. V. Shuleikin:

"Correspondent: Is there experimental verification of the hypothesis that infrasonic oscillations emanated by ocean waves during a powerful storm are capable of generating psychic distress, or even killing a person?

"Shuleikin: No, and cannot be. I spent many years studying infrasound emanated by ocean waves. With increase in wind speed and amplitude, the intensity of the 'voice of the sea' is increased...But even the most powerful hurricane generates infrasound many orders less than the intensity that is dangerous to life."

(No. 43, 1913) wrote: "In spite of the storm...the crew of the Russian steamship set out in dinghies toward the huge, wave-tossed bonfire, and managed to save 102 people. The *Tasr*'s example inspired the crews of other vessels to render more substantial help to the stricken...523 people were saved. All this took place in the midst of a magnitude 9 storm, with waves up to 10 meters tall!"

3.3. A Rift in the Ocean

New York _Times_
October 17, 1989

"It turns out the Americans have been involved in the cover-up of a mystery with global ramifications," Dr. Kang Mao Pang, one of China's leading physicists, announced to reporters. "They hid from society photos of human footprints on the moon for 20 years, and a photo of a human skeleton even longer. What was discovered on the moon is a stunning fact, but Americans apparently feel that no one else in the world has the right to that information."

Doctor Kang declined to reveal how he obtained the photographs and copies of documents stamped "Classified." But who took the photographs is easy to guess from the date on the documents—August 3, 1969. (American astronauts Neil Armstrong and Edwin Aldrin stepped onto the lunar surface on July 21, 1969.)

They say that Armstrong transmitted from the surface of the moon: "Hell, I'd like to know what's going on! Right in front of us, on the other side of the crater, there's two spaceships, huge ones! They're watching us…" And Aldrin, his companion, supposedly shouted into the microphone: "I see individual modules, they're glowing from within." They also said that, while Apollo 13 was circling the moon, the space center in Houston received a radio transmission from astronaut Eugene Cernan: "They're sitting there, like bees on honeycombs."

No official statements regarding these rumors were forthcoming, which generated a lot of false rumors. Some journalists claimed at the time that NASA had made public very little of what the American astronauts actually encountered on the moon, or what they had transmitted into the ether. "Out of consideration for government security," others defended NASA's position. There were also voices claiming that the American astronauts had suffered optical hallucinations on the moon. But photo technology is incapable of hallucinating.

In May of 1969, Thomas Stafford and John Yang twice photographed UFOs in near-lunar orbit, and also while returning to Earth. Aldrin took photographs of two UFOs in the cosmos, which were made public a few years later. One can't say that these were atmospheric aberrations: there is no atmosphere, either on the moon or out in the cosmos.

Imagine for a moment that the moon, which poets so love to praise, is in places hollow within; that beneath its surface lies a network of tunnels connecting craters and enormous caverns. Inside these caves are kept vessels and planes that "vanished" in the Bermuda Triangle, as well as apparatus with whose aid extraterrestrials monitor everything that happens on Earth. It's not hard to conjure up such a picture, since similar things are written about in works of science fiction. However, many have taken as fact information about how huge caverns have been discovered under the moon's surface with the aid of special devices.[31]

I've long puzzled over the question: if these guests from the cosmos really have been visiting our planet for millennia, then why don't they behave like cultured and polite beings? Why not land sometime, say, on the green lawn in front of the White House in Washington? Or Red Square in Moscow? And this should take place in bright, sunny weather, with a large crowd of people. And why not, while they're at it, fling wide the doors of their "flying saucers," so that anyone who wishes to can look inside, the way they do at military air shows? And why shouldn't they, the visitors from the cosmos, or perhaps from a different dimension, let people know the purpose of their visits to Earth?

Of course, one can say that their intentions regarding humanity may be hostile, and that's why they hide. But are these extraterrestrials really hiding? After all, every day, UFOs appear in different corners of the planet, and their pilots even make acquaintance with certain people. At the same time, they exhibit a preference for children, farmers, and housewives, for some reason, but never visit governmental institutions or the publishers of solid newspapers like, for example, the New York *Times* or *Izvestiya*.

---

[31] "Sublunary" caves were discovered in the 1960s. Director of the Pulkovsky Observatory, A. Deich, and his American colleague, Carl Sagan, believed that these caves may harbor conditions conducive to life.

Could it be that the extraterrestrials, foreseeing psychological shock, are gradually training humanity to entertain the idea of their presence on Earth? If this is so, then it's understandable why not all people believe in their existence yet. It also becomes understandable why the American astronauts saw alien flying machines on the dark side of the moon. And if extraterrestrials have bases on our planet, then these bases are naturally located in regions inaccessible to people…

Ray Palmer, publisher of the US magazine *Flying Saucers*, maintains that the inner sphere of our planet is a much more likely source of mysterious luminescence and UFOs than the cosmos. Brinsley L. Trench argues that many anomalous phenomena owe their origin to the underground realm.

It's long been noted that UFOs gravitate toward major geological faults, leading deep into the interior of the Earth. Some of them have been explored in detail by expeditions specially equipped for work in mountainous terrain. But if such a rift is found, say, on the bottom of the ocean, then of course it's inaccessible to humanity…

A geological rift lies in the right corner of the Bermuda Triangle. It's called the Puerto Rican Trench, for the neighboring island of the same name. This is the deepest place in the entire Atlantic Ocean, with a depth of nearly 9 kilometers. To this day, the trench hasn't been completely studied, in spite of thousands of measurements taken by oceanographers. Could it be that this trench, nearly the largest on our planet for length and depth, is the gateway into the underground realm, through which flying saucers reach their bases in the depths of the Earth?

In 1963, American military vessels followed for several days a mysterious object near the coast of Puerto Rico, moving underwater at a speed of 280 km./hour. This was not a submarine, since the maximum speed a contemporary sub could achieve was immeasurably slower. Hydroacoustics indicated that the object sank to a depth of 20,000 feet in a matter of minutes, while specialists performing the analysis insisted that it "must literally belong to another world."

In the winter of 1972, during US Navy maneuvers in the Atlantic, it was observed how "something emerged from the water, having pierced through a 3-meter thickness of ice, and like an enormous silvery bullet, vanished in the sky. The diameter of the object was no less than 12 yards, but the hole in the ice that it left behind was much larger. It swept huge boulders of ice to a height of 20 – 30 yards, while the freezing water in the ice hole was covered with billows of steam, apparently from the red-hot skin of that globe…"

"In the spring of 1986, the Brazilian air force reported the appearance of UFOs in the airspace over the country. The pilot of a reconnaissance plane sent to approach the UFO observed red, green, and white lights, moving at speeds from 250 to 1,500 km/hour, and 'diving' into the Atlantic Ocean."

The geological rift at the bottom of the Atlantic Ocean is not, apparently, the only entrance to the underground realm. If one were to believe ancient Tibetan books, nearly all cave temples in India have underground corridors leading off in different directions. In *Most Ancient England* (1919), Harold Bailey cites the reports of travelers in "the great tunnels" winding underneath a significant part of the African continent.

They say that these tunnels are connected by hidden pathways to the Egyptian pyramids. The Polish traveler Ossendovsky, referring to a story told by a Mongolian lama, tells of a wide network of tunnels under the Himalayas, and "of strange and speedy sources of transportation—devices that circulate in these underground arteries." There are also reports about a many-branched structure of caverns and tunnels, of origins both natural and artificial, under the surface of South America and Europe.

Becoming acquainted with these reports, one gets the impression that our entire planet is enmeshed by a net of underground tunnels and corridors. It's possible that the underground passages, between themselves, connect to many major reservoirs—oceans, seas, lakes—from which UFOs unexpectedly emerge at various points of the Earth's sphere. And indeed, they've been seen more than once in the waters of the Pacific and Indian Oceans, the Mediterranean, and the Red Sea.

3.4. "We Were Terrified…"

A legend has long circulated among mariners, regarding strange rings of light that appear on the surface of oceans and seas from time to time; these regions are said to be the locales where "Flying Dutchmen" are discovered. The spinning rings of light on the water induced fear among superstitious sailors, giving rise long ago to the nickname "sea devil's wheels."

1879. From notes in the ship's log of the English vessel *Welter*: "…Dark night, sky covered with stars. The watchman climbed the mast and confirmed that the illumination was coming from the ocean. It has a characteristic vibration…reminiscent of an enormous, rotating ring. The diameter of this 'ring' is no less than 8 meters."

1880. The steamship *Scheinhein* was making a voyage in the eastern part of the Indian Ocean, when an enormous glowing ring was seen from the upper deck. As soon as the vessel entered the ring, "everything was sunk in a strange, spectral gloom, illuminated rhythmically by running surges of light…"

1976. Pacific Ocean. "Enormous circles of light shone in the water at night. For many successive hours they glided past the boat…Every 2 seconds, light spots break through from somewhere in the depths. Strong, as if a searchlight was turned on there…"

1976. The Soviet oceanographic vessel *Vladimir Vorobyov*, under the command of Captain E. Petrenko, was conducting research in the Arabian Sea. One evening the sailors suddenly noticed that an enormous light patch, rotating clockwise, had spread out in the water around the ship. The echo sounder showed some sort of body lying 20 meters directly under the vessel. At that point the generator powering the trawl winch malfunctioned.

After returning from the expedition, Petrenko told the oceanographic commission of the USSR Academy of Sciences about the strange occurrence:

"All around the vessel, we observed strictly repeating flashes of some sort of unnatural light. The light ran in waves, in the form of eight bent, swiveling light rays, reminiscent of turbine blades. And although we couldn't hear any sounds, the crew members experienced something like pressure on their eardrums. We got the impression that the water was vibrating."

A report was later released, filling in more details of the incident, and appearing in newspapers: "A fiery circle, 150 – 200 meters in diameter, formed in the water, pierced from below by regular impulses of 2 Hertz…The crew members, who had spilled out on deck, said later: 'We were terrified, seized with horror, it seemed that some kind of evil power was at work…'"

That the pulsating light underwater and the malfunctioning of the generator on the *Vladimir Vorobyov* happened simultaneously was not accidental, as shown by two other events that took place not on the sea, but in the sky.

On a February morning in 1983, near the city of Navoi (Uzbekistan), workers at a drilling site observed a flying "apparatus" of large dimensions. In its slightly curved base they could see a small nozzle, pulsing with a scarlet flame.

The workers who saw this UFO got an "eerie feeling." Later some geophysicists, who had been working within 5 kilometers of the drilling site, arrived and said they had also seen that object. As the UFO was flying by, the Navoi hydroelectric station on the Zeravshan River went offline.

Socialist Industry *newspaper*

Moscow,

September 23, 1977

"The citizens of Petrozavodsk were witnesses to an unusual natural phenomenon. On September 20, around four o'clock in the morning, an enormous, bright "star" suddenly flared in the dark sky, sending impulses of light in batches to the ground. This "star" slowly moved toward Petrozavodsk and, sprawling over it in the form of a "jellyfish," hung there, pelting the city with numerous ultra-thin streams of light, creating the impression of torrential rain. After a short time, the light rays abated…

"This phenomenon, according to reports of eyewitnesses, lasted 10 – 12 minutes. The director of the local hydrometeorological observatory, U. Gromov, told a correspondent of [the news service] TASS that the employees of the Karelia region weather service had never before seen anything like it. What generated this manifestation, and what its nature is, remains a mystery."

***Aviation and Cosmonautics Journal* (No. 8, Moscow: 1978)**

"…Engineers, working that night in computer centers located in the region of the sighting, noted major malfunctions in the operation of mainframe computers, whose normal functioning was subsequently fully reestablished."

F. Zigel, science colleague of the Moscow aviation institute:

"…The object had a hard core surrounded, apparently, by glowing plasma…Resemblance to a jellyfish, or umbrella, or parachute…was noted by many eyewitnesses. The light was very bright ('like daylight'). It's characteristic that the light of the UFO was pulsating in nature…"

Professor V. Troitsky, corresponding member of the USSR Academy of Sciences:

"…The Petrozavodsk phenomenon, which was written about in newspapers, to this day hasn't had a sufficiently reasoned explanation. That is, we can't indiscriminately sweep aside the possibility of the existence of UFOs."

Commenting on the mystery at Petrozavodsk, Professor Troitsky at least allows for the possibility of the existence of UFOs. It's unclear, however, what he means by "UFO." Scientifically educated people are so careful in their statements when the subject involves mysterious phenomena, that it's sometimes hard to discern what they really think. But ordinary townspeople spoke more definitively about their observations.

Recorded from the words of eyewitnesses:

"…I saw how some sort of weird, spherical object came down in spirals, and then hovered over the Severnaya Inn. The sphere blinked and gave off a hum. It hung there for 5 – 7 minutes, then the hum intensified, and the sphere flew off in the direction of the lake."

"…I saw a falling star, which I took for a meteorite. But the 'meteorite' didn't fall, instead it stopped and then started moving, quickly growing bigger and taking on the appearance of a body that looked like a blimp, and sharply outlined. The body had a 6-sided form…the sides looked like windows lit up from within. The body moved at a height of 300 – 500 meters, and had a diameter of 12 – 15 meters.

When the body got closer, I got a feeling of anxiety and panic…It was weird, and I ducked to the ground. The length of the body was around 100 meters. As it drew nearer, a glowing white ball flew out from its stern…The ball moved horizontally at first, then came down behind the forest all the way to the ground. This caused the appearance of a bright glow, against which the forest stood out clearly…"

"…Around 4 o'clock in the morning, our team of medics responded to a call on Anokhina Street. About 5 minutes later, a glow appeared over the roof of the opposite house. Then I saw a strange halo and a glowing 'star,' which gave off rays of light that filled a large part of the sky…When one of the rays coming out of the 'star' hit our ambulance, everyone was gripped with a feeling of fear and doom."

From a lecture by ufologist V. Azhazhi:

"…Many residents of Petrozavodsk, both those who later saw the 'object' and those who didn't see it, woke up at 4 o'clock in the morning from some kind of psychological discomfort. Some had nightmares, some had a foreboding feeling of coming catastrophe, and so forth. In any case, people woke up and saw a bright light pouring in through the windows…The 'star' sent down batches of light in pulses…As later established, the intervals were 2 pulses per second, that is, their frequency was 2 Hertz…"

Of course, the feeling of fear may be caused by the spectacle of an unusual manifestation itself. However, the results of a survey of the inhabitants of Petrozavodsk showed that anxiety and fear were also experienced by those who didn't directly observe the UFO. Moreover, the nightmares and sensation of fear manifested in people around 4 o'clock in the morning, when the pulsating light appeared in the sky.

There's nothing surprising here: if the magnetic field emitted by the UFO is capable of "disconnecting" a ship's generator, and even a power station, then what's to be said about the extremely sensitive "electrical grid" of the human brain! By the way, in the *Vladimir Vorobyov* incident, the mysterious underwater object generated not only light but infrasound, since the complex of manifestations—a feeling of fear, pressure on the eardrum, vibration of the water—may be caused by low-frequency infrasonic fluctuations.

If we remember that soon after his encounter with the UFO, Antonio Boas exhibited psychic disturbances resulting, it may be supposed, not just from the traumatic experience, then imagine what might have happened to the crew of the *Vladimir Vorobyov* if the infrasound had been more intense!

3.5. The Voice of Silence

From time to time, I take out from my "extrinsic perceptions" folder materials with descriptions of all kinds of hypotheses, scientific facts, and unusual events, and, like puzzles, I put them together with the hope of constructing a logically significant composition. (This method of combinatorics is very effective in conducting mental experiments). And by the time I got acquainted with the article about Gavro's experiment, my own hypothesis about the natural source of infrasound, the intensity of which causes panic and even poses a danger to life, had already matured. However, let's take it one step at a time…

On July 10, 1958, a team of 10 Canadian mountain climbers arrived at Lituya Bay, situated on the coast of Alaska near the Canadian border. Here they intended to spend the night, but the pilot who delivered them there began to exhibit a puzzling uneasiness.

In spite of good weather, he managed to convince the sportsmen to strike their tents and immediately leave their campsite. They were in the air two hours before the onset of one of the strongest earthquakes recorded in Alaska. The shore, where the mountain climbers' camp had just stood, was engulfed in a gigantic wave.

"Waves near 35 meters tall raced over the place where people had just been standing. A territory the size of Texas and California combined suffered destruction."

Those who have read Jules Verne's novel *The Children of Captain Grant* [also known under the title *In Search of the Castaways*] may remember what happened to Glenarvan and his companions in the Andes. After a long and arduous climb, the travelers settled themselves in an abandoned hut on a mountain plateau. Without warning, there arose a strange cacophony. Everyone rushed outside.

"The bellowing of frightened animals drew nearer. It came from the direction of the mountains, sunk in gloom. What was going on there? And all at once, an avalanche crashed down on the plateau, an avalanche of living creatures, mad with terror. It seemed that the entire plateau shuddered. Hundreds, or maybe thousands, of animals were rushing blindly, generating, in spite of the thin air, a deafening roar…"

What frightened the animals so? After a few hours, Glenarvan and his companions learned the answer: the animals ran because they sensed an impending earthquake. This is not the writer's invention – the fact that many animals can sense an impending earthquake has been known from time immemorial. For example, in an ancient source dated 328 BCE, it says that animals turned fled a few days before an earthquake in Greece.

In 1954, on the eve of an earthquake that devastated Orléansville (Algeria), many domestic animals abandoned their homes. In the same year, the analogous behavior of animals was observed on the eve of a calamitous earthquake in Greece. Those residents who took note of this and left their dwellings in a timely manner were not injured.

In 1976, not long before an earthquake in one of the Italian provinces, cats began to carry their kittens out of houses with frantic urgency. An incident is also described, when a sheepdog dragged a baby out into the yard a few minutes before an earthquake leveled the house.

In the summer of 1923 a bloated, whiskered Pacific cod was found right by the water near Tokyo. Ichthyologists were surprised, since this fish is typically found only at great depths. Two days later, 143 thousand residents of Tokyo became victims of a frightful earthquake. The people who had come upon the whiskered cod must have remembered, on that ill-fated day, the myth about the fish "namadzu," that causes earthquakes by tickling the ocean bottom with its whiskers. An image of that fish has long been pasted on windows as a talisman against underground jolts.

In 1933 a Japanese fisherman in the Odawara region caught an eel, which is typically found at great depths. That very day, three thousand Japanese perished from a strong underground quake.

On November 11, 1963, a deep-water fish measuring 6 meters in length was caught near the island of Nijima, Japan. Since the fishermen had never seen this type of fish before, they took it to the marine laboratory at Aburatsubo. Informed about the rare find, Professor Yasuo Suehiro, an ichthyologist at Tokyo University, noted half-jokingly that, judging by events, an earthquake should be expected any day. Two days later an earthquake actually occurred in the region of Nijima Island. The fact that deep-water fish can have premonitions of earthquakes deserves the most serious investigation, in the opinion of Professor Suehiro.

A number of physical factors are known to precede an earthquake. Apparently, some of them "warn" deep water inhabitants about danger. But which one?

An earthquake occurs as the result of an accumulation of elastic energy in the earth's core, leading at last to a rupture. These elastic forces generate low-frequency mechanical oscillations, in essence, a vibration of the ground. The mechanism that generates earthquakes is easily modeled with the aid of the kind of bow that children play "Indians" with.

If you pull the string tightly, you can feel the bow vibrating in your hands; if you continue to draw the string tighter, the wooden rod will finally snap—an "earthquake." Of course, aquatic inhabitants of the deep don't directly feel the vibrations of the ocean floor, but seismic oscillations pass into the watery environment. These oscillations must be what "warns" deep-water fish of impending danger. After all, vibrations of the ocean bottom often arise long before the initial jolt—hours before, or days, or even weeks. That's where I remembered an article I had stashed in my folder about the "infra-ear" of jellyfish[32]...

It's long been noted that a few hours before an impending storm, jellyfish disappear from the surface of the sea. They go down into the depths, escaping storm waves that may toss them onshore. The factor "warning" jellyfish about an approaching storm turned out to be the "voice of the sea."

On the evolutionary ladder, inhabitants of the ocean bottom are relatively close to the "prehistoric" jellyfish, as conditions for life in the depths have barely changed over the millennia. Therefore, I reasoned, the audio receptor apparatus of deep-water inhabitants must also be "tuned" to infrasound. Thus, the premise of the hypothesis was established: the "voice of the sea" sends jellyfish down, while the "voice of the terrain" sends deep-water fish up, closer to the surface.

---

[32] The "infra-ear" of the jellyfish consists of a stem that ends in a bulb filled with liquid; little crystal "pebbles," or otoliths, float in this liquid, pressing on nerve endings. Infrasound waves, coming from the region of a storm, cause oscillations in the liquid inside the bulb. Nerves convert these mechanical oscillations into electrical impulses, which cause strong "nervous exacerbation" in the jellyfish.

In the first case, the infrasound operates "from the top," while in the second it operates "from below," from the ocean bottom. Neither jellyfish nor fish, of course, "know" that danger is approaching; they simply escape the effect of the infrasound itself.

Life arose in a watery environment. Every motion of a body in the water generates low-frequency oscillations that inform aquatic inhabitants about the approach of an enemy or prey. For this reason, nature ensured that marine organisms developed the ability to sense low-frequency oscillations in the environment.

When organisms "migrated" out of the aquatic environment onto dry land, the situation naturally changed. The enormous variety of natural conditions (mountains, forests, prairies), as well as coexistence with other land animals, stimulated the further development of the hearing organs. Infrasound travels over great distances, but it's not directional. From infrasound alone, it's hard to detect the direction of its source. In terrestrial conditions, therefore, evolution extended the range of sound perception toward higher frequencies.

Although the perception of infrasound on land doesn't carry the same crucial significance for life that it does in an aquatic environment, terrestrial animals haven't completely lost the ability to sense it. Humans, however, cannot hear infrasound. This is explained in part by the fact that the sensory organs that play such an important role in the struggle for animal survival, have, in humans alienated from "nature," functionally diminished somewhat with respect to basically new environmental conditions of habitat. Diminished – but not completely…

At one educational research center, there were experiments involving human isolation. A few volunteers spent six weeks completely cut off from the outside world. All of them had given up their wrist watches in advance. Food and drink were delivered to the subjects through an opening in the wall at irregular intervals, so that they couldn't judge the time of day. A small ceiling lamp illuminated their rooms night and day.

After a few days, the subjects had lost their orientation in time. Except one. It turned out later that this person perceived the sound of an airliner that would pass over the building at a strictly defined time of day. The psychologists' surprise knew no bounds: the plane flew over at a height of several thousand meters, while the room was completely soundproofed!

The psychologists wouldn't have been surprised if they'd known that infrasound is always present in the acoustic spectrum of working plane engines.

As indicated before, the ability to perceive low-frequency sound waves differs among people. For one the lowest boundary of perception may be, say, 30 Hertz, while for another it may be 16 Hertz. There are probably "super-hearers" capable of perceiving sounds at frequencies lower than 16 Hertz. In Japan, which experiences earthquakes more often than most countries, some people insist that they're able to sense disaster ahead of time, similar to people who can sense changes in weather by the pain in their joints…

That people react to infrasound differently depends on individual anatomic and physiological particularities of the vestibular apparatus in the brain's architectonics. This is why, for example, in a region of seismic activity, long before the first subterranean jots, some people will experience anxiety and an unexplained uneasiness, while their neighbors will remain perfectly calm.

Infrasound also generates a disruption of orientation in space. This "seasickness syndrome" testifies to the fact that low sound frequencies have a negative effect on the vestibular apparatus—the organ of balance. Moreover (as shown by the research of psychologists), "stability of orientation is a crucial component of a feeling of confidence, while intense stimulation of the vestibular apparatus generates uncertainty, fear, or panic."

A person experiencing causeless fear may rationalize it as an "intuitive" premonition of disaster. Such a person, overcome by a feeling of anxious uncertainty, like, for example, the pilots Taylor and Kosnar, may back out of an upcoming endeavor.

X

I, the second-year student of the Azerbaijan Medical Institute, shared my idea with Professor Damir Vagidovich Gadzhiyev, head of the biology department at the Azerbaijan medical institute, where I was a sophomore. He suggested I prepare an article.

A month later, my article on the bioprognosis of earthquakes was published in the Baku *Vechyorka [Evening]*. The title of the article—"The PSI-2 Phenomenon"—spoke of my interests: "the PSI phenomenon" was what they called telepathy in those days. It was picked up by all-Union radio for the program, "People, Events, Times," with commentary by Professor Gadzhiyev.

"The Voice of Silence" – that was the title of the article published in the Moscow journal *Znanie – Sila [Knowledge is Power]* (July, 1967). Later, the article was included in the yearly publication *Yevrika [Eureka]*, specializing in the popularization of scientific ideas and discoveries.

Many years later, I acquired two books about inaudible sounds. In them I found a collection of the same evidence as in my publications: animal premonition of impending earthquakes, the "infra-ear" of the jellyfish, Wood's experiment, the investigations of Professor Gavreau, even the citation from Jules Verne, in the same manner I'd presented them in the journal *Znaniye—Sila*. It was, of course, flattering to see my hypothesis in the books of other authors. But I was disappointed that they didn't cite the author of the hypothesis.

3.6. "Humane" Weapon

During the Second World War, near Moscow, in the Volokolamsk region, the Germans were tossing pieces of rail and empty metal barrels full of holes out of airplanes. As they fell, these objects made a noise with an extremely oppressive effect on the psyche. According to Moscow militia-guardsmen, there followed a soul-rending, piercing howl. Everyone plugged their ears, which hardly helped—the roar in one's head lasted all day. What "rent the soul," apparently, was the infrasound generated by the vibrations of the falling pieces of rail and hole-riddled barrels.

I also learned that the biological effects of infrasound were studied in Nazi Germany. The subjects (concentration camp inmates) were experiencing dizziness, excruciating abdominal pain, and breathing difficulties. People's behavior also changed, unconscious fear turned into panic, some people even tried to commit suicide.

The results of studies of the biological effect of low sound frequencies have prompted some to create a "non-lethal" infrasonic weapon. Thus, the newspapers reported about a "novelty" created at the request of the French police. The infrasonic whistle was used to disperse demonstrators. "During the tests of the first model," wrote one French newspaper, "people in the five-mile zone felt the strongest painful vibrations in the whole body. Acoustic weapons were also used in England, during the struggle against riots in Northern Ireland.

"Stopping the development and manufacture of new types of weapons of mass destruction" was the title of an article by a well-known Soviet physicist, Professor A. Fokin, published in the journal *Mir nauki [Science World]* (No. 1, 1978). "The implementation of infrasonic waves at frequencies measuring in the single digits Hertz," the scientist wrote, "actualizes the development of psychotronic weapons: inaudible infrasound generates a condition of anxiety, desperation, terror...

"The influence of a strong infrasonic emission may result in lethal consequences. Death in these cases results from the disruption of the cardio-vascular system with acute changes in blood pressure, or the destruction of blood vessels and internal organs. If we consider the ability of infrasound to penetrate concrete and metallic barriers, then on the basis of this effect, we may expect the appearance of new types of weapons."

The actions and words of Soviet diplomacy, as is well known, frequently diverged. Trumpeting the necessity of forbidding new forms of weaponry, Soviet diplomats could not but know, of course, that their country was involved in the creation of infrasound weapons. This is partially attested to by the testimony of B. Krutikov, an employee at one of the secret laboratories:

"In August of 1991, I was invited to the department of medical technology, where they offered me the chance to take part in the development of the technical documentation of apparatus 'G-1.' They especially stressed the strict deadlines and the necessity of designing an autonomous signal generator, independent of the 'G-1' structure. I made two observations. First—according to the assignment, the generator must ensure work in the range of 10 – 150 Hertz. And the frequencies between 10 – 20 Hertz pertain to infrasonic oscillations, harmful to everything living.

"Second—the autonomy of the generator allows it to be deployed apart from the apparatus 'G-1.' Connect it into a "generator-amplifier-transmitter" system, and you've created an evil 'psy-weapon.' I suggested a change in the technical assignment…They told me it was impossible to exclude infrasonic frequencies from the working range of the generator. They continued to insist on the autonomy of the product. But I categorically refused to work on this project…"

Apparently, Krutikov wasn't the only person in the USSR recruited to work on the development of infrasonic weapons. Thus, during the August 1991 putsch, when Yeltsin's supporters made a living ring around the building in which the Russian government was situated, General Konstantin Kobets announced over internal radio that "there is danger of the implementation of psychotronic generators against the defenders of the White House."

As I became acquainted with these testimonies, I recalled an incident from my student days in Novosibirsk. In the fall of 1966, I was preparing for publication an article on the biological effects of infrasound. Just at that time a conflict was developing on the Soviet-Chinese border—our eastern neighbors decided to annex the border island of Damansky. Later there were heated skirmishes there, in which, according to rumor, many Chinese were laid low. At that point, I got the idea for a "humane" weapon.

If pipes, similar to the one deployed by Robert Wood at the Lyric Theater, were laid along the shoreline, inaudible infrasonic waves generated by them would cause panic in the ranks of the oncoming foe. Not understanding what was going on, the Chinese would run helter-skelter back, but they would remain alive. How is this not a humane weapon! This idea is completely in the spirit of the Chinese warrior philosopher Sun Tzu, who observed in his *Art of War* that "it is useful to kill an opponent, but much more useful to leave him alive."

I shared my idea with a friend who was studying in the radio physical department at NETI[33], and he even tried to make a calculation to correlate the area of the wave emitter with the power of the generator. My studies at the medical institute were coming to an end, and Vyacheslav (my techie friend) was completing his courses at NETI. Absorbed with the subject of infrasound, we decided to devote our futures to it.

I also remembered my innocent conversations with this sympathetic young man. We discussed, among other things, the possibility of "imposing" sleep on someone from a distance. (At that time a device called "Electodream" was widely advertised for people suffering from interrupted sleep.) And what about superimposing the pulses generated by the apparatus over radio waves, for instance, over music?…

Later, I learned from declassified documents that in 1973, in the military unit 71-592 of the city of Novosibirsk, a "Radiodream" device had been created, and testing of it conducted. Could it be that Vyacheslav participated in the creation of this device?

In the summer of 1967, a few days after taking the Hippocratic oath, I received a summons to the military recruitment office. This surprised me, since we "newly-baked" physicians had already completed field exercises while appropriating the title of junior lieutenants in the military service. On Wednesday I appeared at the recruitment office at the appointed time. It wasn't a day for receiving visitors, which puzzled me even more. I opened the door, on which hung a tablet—SPECIAL DEPARTMENT. The person sitting at the

---

[33] NETI is the Novosibirsk Electrotechnical Institute.

desk in a colonel's uniform tore himself away from his newspaper and amiably pointed at a chair. I sat down and waited to see what he'd tell me.

"We know your interests," the colonel finally pronounced. (Of course—a corner of the journal *Znaniye – Sila* with my article on infrasound peeked out from under the papers on his desk.) "We want to offer you…" He fell silent, smoothed his sparse gray hair, coughed a bit. "Work in one of the flight sections outside of Novosibirsk. In general, the work is with healthy people, chiefly with the flight personnel. Research on the influence of harmful factors on the human organism…"

I remained silent, but my head was in turmoil. At that moment I remembered stories about "boot camps," secret laboratories where new weapons were developed. The workers had everything…except the possibility of leaving the "zone," surrounded by guards and barbed wire.

"If you accept the offer," the colonel continued, "Then, with your assignment, you'll receive another star on your epaulets, together with an increase in your physician's salary. Besides that, free housing…" He began to flex his fingers.

"Uniform—also free. Summer and winter. Meals in the officers' mess. The cost is nominal. The food there is delicious, I can attest to it myself." Seeing my confusion, the colonel added:

"Many of your colleagues would be glad of such an offer. You don't need to give an immediate answer. If you decide, give us a call." He handed me a piece of paper with a telephone number…

I didn't phone the recruitment office. The attempt of one scientific research institute to direct me toward research work was deflected by the regional department of health care. I received an assignment to one of the remote areas of the Novosibirsk region—to work off my diploma. For two years. In winter time, you could only get there by helicopter. Incidentally, in the end I didn't regret this assignment: there, in the depths of Siberia, I gained excellent medical experience.

3.7. Are UFOs Rational?

In the attempt to explain the nature of Bermudan phenomena, there's no way we can avoid the subject of UFOs. To begin with, thanks to Berlitz's magic touch, people don't think of the Triangle's riddles in any other way but in connection with visitors from the cosmos, or from a parallel dimension. Secondly, it's impossible to ignore the numerous testimonies that place misfortunes with air and sea vessels in the same regions where UFOs are sighted. For over half a century, passions over this subject now subside, now flare up again, agitating the imagination of world society.

What do unidentified flying objects represent in themselves—natural phenomena, unstudied to date, or the products of an extraterrestrial civilization? Some among my readers will note that, in actuality, one doesn't preclude the other, that in some cases UFOs may be natural phenomena, while in others—the space ships of extraterrestrials.

Well, yes, devil take it, that can also be so!

September 23, 1947

Top secret.

To the Commander of the Army Air Force (Washington 25, District of Columbia)

Attention: Brigadier General George Schulgen, Assistant to the Chief of Staff of the Army Air Force.

"…The phenomenon reported is something real and not visionary or fictitious. There exist objects, disc-like in form, of measurements so significant they don't seem inferior to manmade flying aircraft…

Noteworthy performance characteristics—the rapidity of takeoff, high maneuverability (especially in turns), but also actions worth examining, such as a wish to remain unnoticed and evade meeting with our planes and radar, suggest that some of these objectives have manual, automatic, or remote-automatic control...

Special consideration to the following circumstances:

1. The possibility that these objects are of domestic production—the final product of some top-secret project unknown to Armed Forces headquarters and its main technical management.
2. The absence of material proof, such as: fragments collected after an accident, which would serve as incontrovertible evidence of the existence of such objects.
3. The possibility that some sort of power possesses a type of engine whose principle of operation is unknown to us.

We hereby recommend the following:

By order of Armed Forces headquarters to establish a project under a secret code name, for the purpose of thoroughly investigating these phenomena with the goal of synthesizing all available and relevant data, with subsequent transmission to the various centers of the Army and Navy, the Atomic Energy Commission, NASA...

Pending a special directive, the main technical Department will continue to investigate within the parameters of its capabilities in order to more precisely identify the characteristics of the phenomenon. The most important intelligence will be immediately conveyed to authorities.'

Signed: N. F. Twining, Lieutenant-General.''

From the report of a co-worker at the Massachusetts Institute of Technology (USA, September 29, 1950):

"Our radar intercepted an unknown plane, coming directly from the north at our F-86. It was flying very fast, and I informed the pilots. The image on the screen tore past the F-86 and, cutting across its course, made a sharp bank to the right, turned back, and passed over us. The pilots confirmed that the speed of the object was astounding. It's impossible to believe, and I can't conceive that a human being could withstand such overload on the turns...

"A range of preliminary calculations only confirmed that the object is in many ways completely unusual. Perhaps what perplexes me most is the fact that it produced a very good radar signal; the latter shows an uneven surface and comparatively large dimensions of the object…This episode causes us some bewilderment, and we don't know whether it's permissible to talk about it, or if it's better to keep mum…"

The "cold war" was developing between the USA and USSR. At that time both superpowers took UFOs to be a secret weapon of the enemy. When it became clear that the Russians had nothing to do with UFOs, some American experts began to characterize such objects as "products of an unknown technology." Politicians who fell for this began to seriously carry out plans to entice the extraterrestrials to their side in case of a military conflict. Clearly, the operator of the radar station was advised to keep his mouth shut.

Suspicions about UFOs, as about the secret technologies of the enemy, also arose among Soviet intelligence officers.

Chief of intelligence of the Northern fleet during the 1980s, Captain 1ˢᵗ rank V. Berezhnoi:

"As chief of intelligence for the flotilla, I periodically received reports from eyewitnesses about the discovery of unknown objects…The fact that these objects are observed rather frequently, and the way they behave in areas of nuclear submarine bases, makes me think that they're interested in our technology, and it's no accident that they appear over combat training ranges. There were instances of UFOs hovering over the latest submarine models, even accompanying them in the Barents Sea."

From an interview by a New York *Times* correspondent (January, 1979) with William Spaulding, head of a commission for studying the phenomenon of UFOs, involving 500 American scientists:

"…The CIA would periodically indicate that investigations into UFO sightings had been halted in 1952. However, documents totaling over 1,000 pages, obtained by court order under the Freedom of Information Act, show that the government had been lying to us all these years…After studying the received documents, our group has concluded that UFOs are indeed real, and that the US government has been shown to be dishonest, having instituted a politics of total suppression of information about UFOs…The information was being sent to the CIA, the White House, and the National Safety Council."

In 1947, systematic study of the UFO phenomenon began in the military division of the CIA, within the framework of a project called "Sign." The US

Air Force issued a command to destroy "saucers." However, in attempts to attack them, the planes would blow up. According to testimony leaked to the press, the US Air Force, over a short period of time, had lost 25 pilots.

"From later declassified documents, the US public learned of a memorandum, composed by the military in February 1942 and addressed to President Roosevelt, about the gunning with antiaircraft missiles of a group of UFOs in the skies over Los Angeles. The contents of a report made by the Pentagon to the US government was also made public. The report says, in part: 'We have an enormous number of reports about flying saucers. We take all of this very seriously, as we have already lost a great quantity of people and planes'…"

On September 22, 1976, a telegram from Tehran arrived at the desk of Aleksei Kosygin, Chairman of the Council of Ministers of the USSR. It contained a request from the Iranian government for help in unraveling a situation that occurred over the Iranian capital on the 19th of September.

Two F-4 Phantom fighter jets of the Iranian Air Force had been deployed in order to intercept a brightly shining object in the sky over Tehran. As they neared the object, the pilots discovered that they had lost radio contact. The electronic onboard weapon control system on one of the planes malfunctioned at the very moment when the pilot was about to fire a missile at the object. The apparatus on both planes began functioning again after they had turned aside from the UFO.

The request for help in "unraveling a situation" had been sent to the US government as well. Judging by everything, it appears Tehran had decided that the mysterious object in the sky was a type of new weapon belonging to one of the superpowers.

A year later, a UFO landed near Kuwait City. All telephone and radio connections in the region were interrupted for seven minutes while the UFO remained on Kuwaiti soil. Suddenly, with head-spinning swiftness, the strange object rose vertically. On the morning of November 22, a UFO appeared in the same territory once again, and the same thing was repeated—communication systems stopped functioning. The government of Kuwait appointed a commission to investigate these incidents, but it wasn't able to establish anything concrete…

When it became apparent that the superpowers weren't involved in UFOs, ufological literature began crowding the romance and adventure novels on bookstore shelves. Articles on visitors from the cosmos also began appearing more often in the periodical press.

It's worth mentioning, however, that not all people writing about UFOs identify with ufologists. They are basically representatives of the academic sciences—physics, oceanography, meteorology—who know better than anyone that UFOs are occasionally a manifestation of known natural phenomena. Relevant publications, percentage-wise, are exceptionally few.

Not because there are few skeptics, however—there are plenty of those among scientists. It's just that most of them don't want to be distracted from their main occupations; others don't wish to compromise themselves by participation in what they feel to be discussions unworthy of serious scientists. It therefore happens that the arguments of opponents of the idea about alien spacecraft are drowned in a multi-voiced chorus of ufologists…

3.8. On the Benefit of Comparisons

UFOs are credited with various, both purely external features and physical properties. And by mere observations of eyewitnesses, it is not an easy task to determine what these objects are—man-made products or natural phenomena. However, I was prompted to turn to ufology by the long-noted similarity of some Bermuda phenomena with the physical features inherent in "flying saucers". For example, in the Bermuda Triangle, communication systems and electronic equipment on ships and aircraft sometimes go wrong, or even completely fail…

"It was probably between 21:30 and 22:00 by the clock. I don't know for sure, because I didn't have a watch. I was working on the tractor…with my brother. Suddenly we saw a source of light, so bright that it hurt our eyes. The light was coming from a huge, round object that looked like the wheel of a car. The light was bright red, it lit up a large area. I suggested to my brother that we go take a look to see what it was. But he didn't want to.

"Then I went alone. When I got close to the object, it suddenly started to move, and rolled with incredible speed to the other side of the field, where it stopped again. I ran after it, but the same thing happened. Now it came back to its previous place. I made no fewer than 20 attempts to get close to it, but without results. I got offended, and went back to my brother…"

And here's another testimony: "…The author of the letter saw how, after a hit of linear lightning in the plowed field, multicolored splashes of flame seemed to run along it, so it looked as though the soil 'caught fire.' After three to five seconds, the "splashes" gathered into a circle, forming a huge sphere of white light…The sphere tore away from the tillage, rose two meters from the ground, and began moving along it. Its surface looked like it was boiling…The sphere moved toward the wall of the yard, near which the author of the letter and his father were standing. Two-and-a-half to three meters from the wall, it sharply changed course by 90 degrees…"

The stories cited here resemble each other, though the second one doesn't appear as intriguing—it doesn't contain details such as "the wheel of a car." The thing is, the first storyteller presents the "round object" as the flying apparatus of extraterrestrials, while the second is describing the behavior of ball lightning. The first story belongs to our friend, the Brazilian farmer Antonio Boas, while the second belongs to the body of scientific literature about ball lightning.

The story told in the farmer's words by the journalist Martins from Rio de Janeiro quickly gained world fame. But academic literature is only read by a narrow circle of specialists, and the evidence contained in it seldom reaches a wide public. I'd like, at least in part, to fill in this blank and acquaint the reader with some of the peculiarities of ball lightning. We'll let Antonio continue his story, and in the relevant parts of his report I'll interrupt with citations from the scientific literature.

"…For a couple of minutes, the shining wheel stayed still. From time to time it seemed that there were rays pointing in different directions coming out of it. *("The surface of ball lightning may have the appearance of solid flesh. Very often it's uneven and continually distorting. Sometimes this gives the impression of small tongues of flame.")*[34]

Then everything suddenly vanished, as though the light had been extinguished. *("Ball lightning may extinguish spontaneously and completely unexpectedly for the observer.")* I'm not quite sure if it was all really like that. It could be that I turned away for a moment, and right at that time it quickly rose and flew away…

"On the next day I was alone, plowing the same field. It was a cold night, and the clear sky was sprinkled with stars. Right at one o'clock in the morning, I saw a red star that looked just like the big, bright stars. *("Ball lightning manifests not only in periods of heightened thunderstorm activity. There are 134 observations in which ball lightning manifested in clear weather.")* But I noticed right away that it wasn't a star at all, because it was growing bigger and seemed to be getting closer.

"In a few moments, it turned into a shining, egg-shaped object *(In a survey conducted by NASA, out of 112 events, in 98 cases a spherical form was recorded, in 9 cases an ellipsoid form, and only in 3 cases a ring-like form was*

---

[34] Here and below, citations in cursive are from the monograph, *On the Physical Nature of Ball Lightning* by I. P. Stakhanov (Moscow, 1979).

reported"), rushing at me so fast that it was above the tractor before I had time to think what I should do…

"While I wavered, not knowing what decision to make, the object moved slightly and once again stopped approximately with 10 – 15 meters of the tractor. Then it slowly lowered to the ground. (Ball lightning *"either falls to the ground, or stops near the surface of the ground, where the power of gravity equilibrates."*); finally, I could see that it was an unusual, almost round machine, with small red openings.

"A great red searchlight was shining in my face, blinding me, when the object came down. Now I could accurately see the form of the machine. It looked like an elongated egg with three spikes in front. I couldn't tell what color they were because they were drowned by the red light; something on top, also red, was spinning very fast…*(Spinning of ball lightning is recorded in approximately 30 percent of events.)"*

"The main details I noticed later, of course, because at first I was greatly agitated…I lost what was left of my self-control, when, a few meters off the ground, three metal nozzles, as in a tripod, appeared from the lower part of the object. These were metallic legs…I didn't want to wait any longer. The tractor was running this whole time. I gave it gas, turned it in the opposite direction, and tried to get away. But after a couple of meters the engine stopped and the headlights went out.

"I couldn't understand the reason for this, because the engine was running, and the headlights worked. *("Close to an object, ball lightning causes in it a significant induced charge.")* The engine wouldn't start. Then I jumped out of the tractor and took off running. But it was too late, because after a few steps someone grabbed me by the hand. It turned out to be a small, oddly dressed creature, about my shoulder height. I turned to it in utter despair, and punched it, which made it lose its balance…I wanted to run again, but was immediately caught by three other creatures, just as strange…They were dragging me to the machine, which stood 10 meters off the ground on the metallic legs described earlier…"

I won't insist that the story told by Antonio Boas is a total fabrication. However, it cannot be ruled out that ordinary ball lightning excited the imagination of a simple lad who wanted a free ride to the capital to become world famous. Or maybe he suffered from a mental illness? In any case, Juan Martins had a suspicion of mental trouble in relation to the Brazilian farmer.

Here is what the journalist writes about this: "…After an exciting event that happened to him, young Boas returned home completely exhausted and slept after that for almost a whole day. Already in the next and subsequent nights, he began to experience insomnia…He woke up in fright, screaming…As the day came on, he ran back and forth and smoked cigarette after cigarette. When he was thirsty, he managed to drink only a cup of coffee, after which he became ill, and the state of nausea, as well as headache, continued throughout the day…"

Dr. Fontes, who examined Antonio Boas, writes in his report: "…Antonio Boas did not suffer from any mental disorders. Neither was he a flawed person, that is, a pathological person who uses every opportunity to be in the spotlight."

Below, I cite one more story published by a Russian-language journal in the USA.

"In the middle of the past week, Yana Palatnik, a resident of Brooklyn (New York) called the office and told the following: On Saturday, January 27, she and her husband David were returning by auto from Baltimore…The sky that evening was clear and starry. When the Palatniks' car passed by Exit 9…Yana suddenly saw a large, fiery ball above the horizon to her right. 'Look,' she said to her husband, 'what a beautiful sunset…' David said, 'Wait, what sunset are you talking about, the sun set several hours ago…'

"To their surprise, the fiery ball didn't change its position, didn't hide below the horizon, but suddenly began to move parallel to their course toward New York. Sometimes the ball would stop, sometimes it would rise slightly, then lower, but to the Palatniks it seemed that it was following automobiles. *("Coming down, ball lightning…alternately rises, then lowers over the surface of the ground while moving horizontally, following the form of the terrain…")*

"Over the course of the next 20 minutes, the 'object' changed color several times, as it alternately drew away from or got closer to the highway. *("The duration of the life of ball lightning changes within wide parameters, partly from inevitable errors of observation.")* They saw how the ball, by now a bluish color, sharply went off in the direction of a bridge…*("Evaluations of the color of ball lightning have a subjective character. This has to do with the variation in color and tint that one and the same ball lightning may possess.")*

"The most amazing thing was that it would suddenly stop in mid-flight and change course, which clearly told all witnesses of this spectacle: before them was an unearthly 'creation.' Everything the ball did in the course of its flight, especially its stops, categorically contradicted the formal laws of physics. *("The charge of ball lightning changes in the process of its motion... consequently, the character of its motion may change unexpectedly...Its motions seem strange and hard to describe. Just the fact of its independent movement in space, minus any connection to conductors that might supply it with energy, naturally causes wonder in eyewitnesses.")"*

This is how the editors of the journal commented on Yana Palatnik's report: "Of course, it's hard to comment on our reader's story. We can only add it to the enormous quantity of testimonies by people from all parts of the world who, like the Palatnik couple, saw a UFO in the sky or on the ground. It's still hard to guess, today, when humankind will be able to figure out who is sending these singular guests to our Earth, and why."

X

The objects of observation of both Antonio and the Palatniks, of course, were fireballs. But the ufologist will not fail to oppose the "skeptics" with an irresistible argument: the technical devices of an alien civilization can also turn off the tractor engine, make strange movements…

3.9. Radio Physics Department

I recall my presentation at the NETI. Shortly after my article about "bioprediction" of earthquakes appeared in the journal "Znanie sila" I received a letter from this educational institution. In it was a request to connect with the head of the radio physics department, Professor N. I. Kabanov. I phoned the next day. The Professor suggested making a report to his department, confirming that his student (Vyacheslav) had told him about my hypothesis.

I agreed, though I couldn't figure out what it was about my work that could have interested Kabanov, whose spheres of interest were the ionosphere and radio waves. Truly, what did electromagnetism and infrasound have in common? (In later years, when I racked my brain over the mysteries of the Bermuda Triangle, more than once, like a spark in the subconscious, the thought would flash by: these two physical factors are inherent in UFOs—magnetic fields and infrasound. But at that point I didn't yet possess the requisite knowledge to think through the connection.)

The meeting was held in Professor Kabanov's office, where there were about thirty listeners—students and several employees of the Department of Radiophysics. I could see the bewilderment on the boys' faces. They, accustomed to lectures by professors with scientific titles, were now being offered to listen to a medical student, who, it was announced, would share his hypothesis. His hypothesis? Well, well…

As an introduction, I spoke briefly about the biological effects of infrasound (experiments by Robert Wood, Vladimir Gavro). Since my listeners were technicians, I made an emphasis on the possibility of creating technical devices. First of all, the bionic aspect. In the 1960s, when bionics were in fashion ("the daughter of cybernetics"), much was said about the possibility of copying living nature and bringing its principles into technology.

When I came to the deduction that animals are "warned" about earthquakes through infrasound, I thought of the possibility of creating an apparatus that would copy the audio reception mechanism of deep-water inhabitants. This device, I thought, being lowered down a deep shaft or to the ocean bottom, would be able to warn of impending earthquakes.

The second aspect is the one dealing with health: defense against the harmful effects of infrasound. Sounds form at various frequencies, on city streets due to multiple forms of transport, and at businesses due to technical complexes; infrasound occupies a significant place in this spectrum. Blending together, the sounds generate noise. Noise is mitigated against, but infrasound isn't considered—after all, people can't hear it.

Meantime, frequencies in the infrasonic range coincide with the basic electrical rhythms of the brain, in particular the alpha rhythm, characterized by frequencies of 8 – 13 Hertz. In other words, a "collision" of biological and technical frequencies takes place in the human brain, which apparently reflects on a person's psychic condition.

It's possible that infrasound in many ways causes nervous disorders, such as neurasthenia and depression, in dwellers of large cities. It seems possible that devices could be developed, a species of "silencer" that would generate infrasound on the same acoustic spectrum as the technical objects, only in reversed phase. By this method, it would be possible to almost completely neutralize the effect of the negative factors.

Talking about the psychological effects of infrasonic waves, I could not, of course, not mention the tragic event with the Nineteen Section planes. I must say that here I had to endure several unpleasant moments. As I had anticipated, there were some protesting cries. Mainly the objections were that one should not take for granted all kinds of hokum, which the western press feeds to its readers.

I answered that these were not stories, that the US Air Force had published the voluminous report of the investigation of the accident.

"Excuse me," the long-necked student raised his hand.

"Is it correct to base a scientific hypothesis on questionable publications?" he turned to the audience with an ironic smile. After a significant pause, he continued: "Let's assume, however, that the story about the disappearance of five American planes took place in reality. You claim (the guy turned to me) that Lieutenant Taylor's unwillingness to take part in the training flight was due to the psychological effect of the infrasonic waves…

"I am not asserting," I interrupted my opponent. "I suggest the most probable reason. Infrasonic waves induce an unmotivated feeling of fear, panic. This is a proven fact."

"You can't argue with the fact," the future physicist nodded his head. "But you, colleague, treated this fact too freely—you drew it to your speculative hypothesis. Judge for yourself (he looked around the audience again). An hour before leaving, Taylor was at the airfield in Miami, hundreds of kilometers from the epicenter of supposed earthquake. An underwater earthquake! Meanwhile, any schoolboy knows: as sound waves move away from the source, their intensity diminishes…"

My opponent threw up his hands and stuck out his lower lip.

"Did the seismographs register an earthquake that day?" One of the teachers asked.

"There's nothing about that in the commission's report," I hesitated. And hurriedly added: "You see, there are different types of infrasonic waves, and different models of seismographs…"

For a short time there was an awkward silence. I was aware of the vulnerability of my hypothesis. But confidence in its validity allowed me to quickly pull myself together.

"It's not the earthquake itself," I said, raising my voice. And clearly, almost syllable by syllable, I formulated the main idea once again: Infrasounds arise in the period of preparation for the earthquake, a few hours, or a few days before it. The stronger is the seismic tension, the more intense is the infrasound…

"Have you read the relevant publications of geologists?" asked an elderly lady sitting at the door. Maybe there were studies…well, that infrasonic sounds are registered during seismic stress, long before the earthquake?

I did not find such information in the literature…

Professor Kabanov came to the rescue.

"Let's be lenient to our guest," he said as he rose from his chair. "Every hypothesis raises questions—that's natural. In my opinion, Boris' idea has a right to exist. For!" The professor held up his index finger. "As it was rightly said, you can't argue with the fact: many animals really feel the approaching earthquake. Hopefully, in the near future the hypothesis will be confirmed by seismologists…"

The "near future" had to wait many years. One day, "walking" on the Internet, I accidentally came across a message about the International Symposium on Problems of Geophysics, which was held in Kazakhstan in September 2016. Among the published materials was an article by researchers of the Ionosphere Institute of the National Center for Space Research and Technology[35].

I will limit myself to a few quotes from the published article.

"The scientific literature discusses the issue of acoustic waves, which, when seismic activity increases, propagate over considerable distances and are recorded in the surface atmosphere before the main shock.

"During the earthquake of August 15, 2014 (in Kazakhstan – B.O.) the infrasound complex of the Institute of Ionosphere registered the appearance of infrasonic waves in the surface atmosphere on the eve of the earthquake."

"…About 3h 10 min before the earthquake there were positive bursts of atmospheric electricity. Notes that perturbations in atmospheric infrasound and electricity appeared before the earthquake at about the same time."

"A study on the detection…in the surface atmosphere of infrasonic waves and atmospheric electricity during periods of activation of seismic processes has been carried out."[36]

---

[35] The publication came out two years after the publication of the third edition of this book in Russian "Psychosis of the Planet Earth" (Moscow, AST, 2014)

[36] Salikhov et al. Experimental revealing of the response of lithospheric processes in background atmosphere in the period of intensification of seismic processes (in Russian).

3.10. "Illegal" Storms

Boston,
December 29, 1988

"Seven seamen on the container ship *Lloyd Bermuda* vanished without any news after their vessel sank on Wednesday evening, 160 miles east of New Jersey.

"The vessel, which makes regular trips on the route between New York and the Bermuda Islands, encountered an extremely brutal storm. Icy, torrential rain was accompanied by hurricane-force winds…"

What can we say about this newspaper article? Nothing exceptional, the usual unfortunate incident. You ask: why, then, do weather services exist? It's a reasonable question—after all, today's weather services can predict pretty accurately where storms and hurricanes will form. Pertinent information fills the airwaves, while vessels and airplanes rush to avoid the dangerous area.

So why wasn't the *Lloyd Bermuda* warned of the advancing storm? Apparently, because the weather in the Bermuda Triangle is often unpredictable. As noted by the American journalist Michelmore, "in the Bermuda Triangle, all troubles arise suddenly."

"Just as "suddenly," a different brutal storm came down on the Soviet oceanographic vessel *Vityaz'*, during an investigative expedition in the Bermuda Triangle in 1982. As one member of the expedition tells it: "…From nearly complete calm to magnitude six! And only in the space of a few hours! By nightfall the agitation had reached magnitude 9, and the wind—magnitude 11. A serious storm…In the cabins, refrigerators were turned out of their sockets, dishes shattered in the galley, on deck the wind howled, vomiting tore the jowls apart…

"The storm went on for three days. The weather forecasters could only spread their hands. We received radio transmissions of two synoptic maps, one from Europe, the other from America. Both maps, based on readings from numerous weather stations, weather ships, and satellite data, unanimously claimed that at the present moment everything outside our ship was smoothness and calm. But in actuality, around us the ocean is howling and standing on end. Where did this illegal, unaccounted storm come from?"

"All incidents in the Bermuda Triangle conform to a single exhaustive explanation. The main cause of catastrophes in this area is due to the meteorological conditions prevailing there, with characteristic sudden, acute deterioration of weather over relatively small areas," says Dr. Root, a meteorologist at the University of Miami. However, if one thinks about it, his explanation is precisely not exhaustive: the question of why, in this region of the Atlantic, the weather sometimes changes so drastically it can't even be predicted, remains unanswered.

Back in the 17th century, the German astronomer Johannes Kepler proposed that weather, in many respects, is dependent on cosmic factors. His contemporaries didn't believe him. Today, indication of cosmic factors pertaining to weather may be found in the academic literature: "Disturbances, elicited in the magnetosphere by solar wind, exert a direct influence on weather, and are capable of bringing on catastrophic hurricanes."

So, magnetic storms! Since the timing of the onset of magnetic storms may frequently be predicted, it wouldn't be that hard, one would think, to prognosticate drastic changes in weather. But that's just it, hurricanes don't always occur on days when the sun is active. And if, in the array of factors influencing weather in the North Atlantic, a major role is played by magnetic disturbances, the cause of the latter isn't always due to solar activity; sometimes they have a different source, which is apparently located in the ocean depths. And actually, many earthquakes are also accompanied by strong magnetic storms…

"Since the beginning of the 21st century there has been a steady increase in the number and strength of tropical cyclones, or hurricanes (called typhoons in the Pacific Ocean and Southeast Asia). Some scientists rushed to blame this on global warming. Subsequently, however, it was noted that many hurricanes arose in the ocean depths, where there occurred earthquakes or movements of the Earth's crust. With the aid of satellites, it was possible to study the paths of hurricanes: "the trajectories of their movement coincided with the lines of major geological fault lines"."

X

At the end of the 1950s, the above mentioned Soviet oceanographer Vasily Shuleikin wrote: "There needs to be a deep investigation—in the literal sense of the word—of the most interesting region of the Atlantic Ocean, in the shortest distance between the continents of Africa and South America…" The motive for such a recommendation was provided by the results of investigations showing that in the indicated region of the Atlantic (which includes a majority of the Bermuda Triangle) there flow electrical currents of unusually strong density.

"It's hard to understand," Shuleikin wrote afterwards, "why investigators, having long ago discovered electrical currents in the hard crust of the Earth—so-called earthly, or telluric currents—have for so long not been interested to learn whether there pass through the waters of oceans and inland seas significantly large currents possessing major conductivity."

The scholar took up the goal of determining to what extent telluric currents in the ocean depths contribute to the generation of uncommonly large magnetic declination. With that, investigations showed that "in the most interesting region" of the Atlantic the sizes of magnetic declination are subjected to unusually acute fluctuations.

"How is it possible not to become alarmed?" exclaims Shuleikin. "How is it possible not to become absorbed in figuring out the role of the ocean in that field of geophysics where its role until this very day many have attempted to deny?"

Further investigations revealed that the density of currents in other major reservoirs of the planet is not enough for generating any significant magnetic declination.

Truly, how could you not become alarmed? After all, in the waters of the Bermuda Triangle there sporadically arise electrical currents of unusually high density, atypical for other regions of the world's oceans! And if these currents "disturb" the natural magnetic field, then it's understandable why navigational systems occasionally glitch in the Triangle.

They glitch so badly that navigators of marine and air vessels, traversing this region of the Atlantic, confirm their position using ancient astronomical methods of navigation—by the stars, with the aid of a sexton. Magnetic storms, spawned by telluric currents, don't envelop the Earth's sphere as a whole, but have a local, so to speak regional, character, and for this reason aren't registered by ionospheric-magnetic stations.

From scientific literature:

In 50% of cases magnetic storms are associated with processes occurring in the Earth's interior. It used to be believed that magnetic disturbances were caused exclusively by solar flares.

3.11. Celestial Blinds

In December of 1970, Bruce Gernon experienced something incredible. Together with two passengers, he flew out from the island of Bimini, headed for Miami in a small private plane. At a height of 3,500 meters he noticed a cloud with an unusual ellipsoid shape.

"It looked innocent and immobile," Gernon recounted later. "I decided to fly over it. I began to gain altitude, but soon discovered that the cloud was rising after me. At times I was able to get farther away from the cloud, but it caught up with me every time, and covered me…I began to think perhaps I should turn back, when the plane unexpectedly left the cloud, and a perfectly clear sky stretched before me.

"When I glanced back, I saw that the cloud was incredibly quickly growing to gigantic proportions, and had taken the shape of a half-moon. Its upper part reached a height of two kilometers, while the lower part touched the surface of the ocean. My plane seemed to be inside a cloudy half-moon…"

Gernon drove his plane forward at maximum speed, but what happened during the next minutes threw him into consternation: the "half-moon" began to quickly bend, and soon the plane was in the center of a cloud that had somehow rolled itself up into a bagel. The cloud was frighteningly unusual, and Gernon started looking for a way out of it. Finally, he saw an opening through which he could pop out. But the opening was rapidly narrowing, and soon became something like a tunnel. Gernon darted into the "tunnel," but its walls also began to narrow.

"I had the impression," Gernon recalls, "as though I'm flying through the barrel of a gigantic gun, about a kilometer and a half long; through the opening on the other side I could see the Florida coast." The clearly delineated walls of the "tunnel" continued to narrow, emanating a shimmering, whitish light…

Soon the whole area was enveloped in some sort of greenish-whitish fog. Nothing could be seen, only dense fog. At this point, Gernon and his passengers experienced a state of weightlessness, and lost their sense of orientation. Gernon wanted to determine their location, but the navigational instruments weren't functioning, and he was unable to contact ground services by radio. Suddenly, wide, horizontal slats opened up in the fog; you could look through them, as though through blinds.

The slats opened wider and wider, and finally Gernon could see the shoreline of Miami through them. When the plane landed at Palm Beach, Gernon discovered that the entire flight had taken only 45 minutes, instead of the usual 75. Half an hour had somehow "trickled away"…

This story was discussed in a series of publications about the Bermuda Triangle; it contains within itself a whole array of riddles. Correspondingly, an array of conjectures was expressed, beginning with the natural (the plane ended up in a developing thundercloud), and ending with a hypothesis about the intrigues of extraterrestrials. Opinions were also voiced that Gernon and his passengers had possibly experienced hallucinations during the flight – what else could celestial blinds be!

It's doubtful that the cloud was a thundercloud: it was located at a height of over three kilometers, when thunderclouds form much lower. Hallucinations? Of course, there can be mass visual hallucinations, when many people "see" them at once. But there are other, completely objective, circumstances in Gernon's narrative: the nonfunctioning of navigational instruments, the cutting off of radio contact, the change in the progress of time…

X

Back in 1924, the Soviet geophysicist E.A. Chernyavsky turned his attention to the agitation of the geoelectrical field on the eve of an earthquake. In the summer he came to Jalal-Abad (Kirgizia) as part of an expedition researching atmospheric electricity in field conditions. "…We were struck by the unusual behavior of the instruments," wrote the scientist.

"The apparatus obviously indicated that an 'electrical storm' with an extremely high potential was raging in the atmosphere. How high, we weren't able to measure, because the arrow of the apparatus immediately went beyond the range of the dial. Two hours later, the earth split. Then I thought: perhaps the earthquake was after all the reason for the anomalous condition of the atmospheric electricity?"

Years later, observations by geophysicists showed that 4-10 days before an earthquake, circular and linear cloud anomalies (CLA) form in the atmosphere along tectonic fault lines. Such clouds were called seismic clouds. Obviously, CLA are ionized air masses formed under the influence of geophysical processes. Occupying huge spaces in the near-ground atmosphere, these inherently electric fields "discharge batteries, disable electric equipment, radio equipment and engines, through current leakages and short circuits in electrical circuits, on ships and planes."

X

In 1974, the US Coast Guard cutter *Hollyhock* was coursing in the western Atlantic not far from the Florida shore. Suddenly the duty officer, Lieutenant Bissman, saw an enormous expanse of dry land resembling an island on his radar screen, situated at a distance of 18 kilometers. The officer couldn't believe his eyes: there weren't supposed to be any islands in this vicinity! When the *Hollyhock* approached the site, nothing was there.

The *civilizants* explain this incident thus: many thousands of years ago, there actually was an island at this location, and the radar picked up signals from the distant past. Lieutenant Bissman later lamented that his message had been interpreted incorrectly. He clarified that occasional radar malfunctions are not that rare, occurring due to technical interference or atypical weather conditions.

It looks as though "atypical weather conditions" in this case generated atmospheric electricity. It happens sometimes that ionized air masses aren't visible to the naked eye, but appear with corresponding contours on a radar screen; seen from a distance, the contours blend with the horizon and may be taken for dry land.

An incident with an Eastern Airlines Boeing 727 may be similarly explained. Entering a light mist, the plane disappeared from Miami airport radar screens for 10 minutes; after landing, it was discovered that the timepieces of the passengers and in the cockpit were 10 minutes behind. Does such a coincidence mean that the plane had been in a different, "parallel" dimension for 10 minutes?

Don't be too quick to answer this question in the affirmative, reader – after all, the light mist in which the plane spent 10 minutes may have been precisely a seismic cloud. Similar to polar flares, ionized air masses present ideal screens for reflecting the electromagnetic impulses by which radar "feels" objects in surrounding space. Understandably, under such conditions a plane won't appear on the radar screen.

A similar phenomenon was observed in the Bermuda Triangle on April 3, 1974. The US Coast Guard cutter *Dakota* radioed that a pleasure craft, the *Elizabeth II*, drifting nearby, had suddenly vanished from the radar screen; at the same time, sailors on the *Dakota*'s deck continued to see it with the naked eye.

"But what about the loss of time?" the reader asks. "Half an hour somehow 'trickled away' from the pilot Gernon…The Boeing 727 passengers lost 10 minutes…"

Here it's appropriate to note that Lawrence Kusche includes time among the factors causing the tragedy of Flight 19. "The report says than not a single one of the planes had a clock," writes the former pilot, "though it's not known whether the pilots had wrist watches. Because Taylor repeatedly asked about the time, it's clear that he didn't have a watch," Kusche concludes. To me this seems impossible, since training flights are not conducted without accounting for time…

You probably know, esteemed reader, how some irresponsible people "economize" on electricity? They place a horseshoe magnet on the electric meter, causing the metallic disc of the measuring device to stop rotating. No one, naturally, thinks that this saves energy. No, the energy is consumed as usual, but the counter isn't functioning, so money spent on electricity is "economized."

This begs for an analogy, doesn't it? On planes flying through an electric field (or near the source of a magnetic field), compasses begin to "lie," or completely stop functioning; apparently, the hands of a timepiece (which in the last century was not shielded from the effects of a magnetic field) may also be blocked.[37] Therefore, in the cited instances it's better to speak not about time as a property of physical space, but of clocks that measure time.

---

[37] In June of 1965, a UFO maneuvered for a duration of 45 minutes over the airport at Santa Maria, California. When the object hovered over the airfield, the hands of the

"But what about the celestial blinds?" the reader persists. "What are those, a hallucination? Or a display of extraterrestrial technology?"

Neither one, nor the other. This is yet another Bermudan phenomenon—a rare, but completely natural phenomenon which, by the way, may be observed in other seismologically active regions of the planet. Here, first of all, it's worthwhile to turn our attention to the weightlessness and temporary loss of orientation experienced by Gernon and his passengers—a clear indication of the irritation of the vestibular apparatus…

And so, infrasound! A dense cloud, in which lateral infrasonic waves result in compression and rarefaction, ceases to be homogeneous—decompaction sites will become more transparent. There you have your "celestial blinds."[38]

electromagnetic clocks in the control tower stopped working. When the UFO flew off, after hovering near the tower for 10 minutes, the clocks began to function normally. The author found a description of a similar event in one of the Russian newspapers. The article cites a letter from a reader who had ball lightning fly into his house. It didn't cause any harm. But all the clocks in the house were lagging by 15 minutes. This was discovered when the correct time was announced on TV.

[38] There is a hypothesis according to which the ionization of air masses also occurs under the influence of infrasonic waves.

3.12. A Subterranean Jolt

During the Gobi-Altay earthquake of 1957, several Mongolian shepherds lost consciousness. This occurred, according to their companions, even prior to the main jolt that resulted in destruction. As the sufferers themselves related afterwards, they sensed neither a noise nor a jolt. The fact that only some of the shepherds lost consciousness is apparently explained by the differing reactions of people to the effects of low-frequency sound waves.

If the infrasound had been more intense, however, it may be surmised that there would have been nobody to tell about the event. Thus, after a strong earthquake in Peru, many citizens perished, as is written, for unknown reasons. That is, they didn't die under the ruins of buildings, or from fires, but for causes that were invisible.

A major role in similar incidents is played by so-called shock waves, generated by powerful seismic jolts.[39] The strength of shock waves may be judged by the occurrence of a catastrophic earthquake in Armenia (magnitude 6.8) on December 7, 1988. Known as the Spitak (or Leninakan) earthquake, in half a minute it destroyed almost the entire northern half of the republic. "Layers of earth, people, houses, buses, rose into the air…a heavy military tank that happened to be sitting over a deep fault was lifted into the air."[40]

Similarly, shock waves manifest on watery expanses. According to eyewitnesses, the consequences of the 1958 Alaska earthquake were "monstrous" even on the ocean. "One fisherman, who was a few miles from shore, later related that his boat was literally tossed upward by strong explosions. According to another fisherman, his boat jumped approximately

---

[39] A shock wave is a sharp, almost instantaneous, change in environmental parameters: density, pressure, speed.

[40] The focus of the Spitak earthquake was located in the junction zone of the Arabian and Eurasian lithospheric plates, at the crossroads of two major geological faults.

four meters high. Radio operators on ships were intercepting pleas for help in the ether."

The earthquake occurred on August 8, 1868. That day a US Navy gunship, the *Wateree*, was anchored in the port of Arica. An officer of the *Wateree*, Lieutenant Billings, left an account of this natural disaster.

"...Around 4 o'clock in the afternoon the captain and I were sitting in the cabin. All at once, we jumped up: the vessel was vibrating as though the anchors were being lowered, and the chain was rattling in the lock. Knowing that this couldn't be, we ran out on the captain's bridge...

"Before our startled eyes the hills, it seemed, were swaying, while the surface of the ground undulated, as though just underneath it short, irregular waves were running in a chaotically agitated sea...

"Part of the night had passed, when the watchman's cry rang out, warning of the approach of a tidal wave...[41] *With a deafening roar, the wave swallowed our ship, engulfing it in masses of water and sand. We remained underwater for some time, which to us seemed like forever. Then, its entire framework creaking dully, our sturdy old* Wateree *made it to the surface together with all its crew...*

"With frightening swiftness, the flow carried our ship with it, but soon stopped. After waiting a bit, we finally lowered a lantern over the side. It turned out we were firmly settled on a bank...We were on dry land, 5 kilometers from our berth, and 3 kilometers from the shore. If the wave had carried us 200 feet further, the ship would have smashed against a sheer mountainside...

"The ground continued to shudder...but not a single jolt reached the strength and duration of the first. Still, some of them were sufficiently strong, and shook the Wateree *with such force that it would tremble like a kettle full of boiling water. Forced to abandon the ship at such moments, we would climb up to a plateau 200 meters higher..."*

You may ask why the mariners intermittently abandoned their ship. After all, the plateau to which they climbed was also shaking. It shook at the same

---

[41] A "tidal" wave has nothing to do with tides. It's formed as the result on an underwater earthquake, or the eruption of an underwater volcano. Vessels caught in such a wave experience a short, sharp jolt. To the sailors it feels as though they've run aground on a rock formation.

frequency as the body of the ship. As pointed out before, the reason lay in the construction of the vessel, a kind of sink-resonator that amplifies the effect of infrasonic oscillations upon the organism. And when the intensity of vibrations escalated, the sailors, tormented by unbearable pain, vertigo, and nausea, were forced to leave the ship and scramble to the plateau. And what could happen if the ship were in the open ocean?

"…'This is like the stroke of a dagger! Hurry and help!' called the anguished voice over the radio. 'Hurry, we're goners!' And then the cries for help emanating from the *Raifuku Maru* died away in the placid calmness of the ocean. Crews of other ships in the area of the Bermuda Triangle were perplexed: what could cause a vessel to cry out for help in such quiet weather?"

Having presented the Legend, Kusche acquaints the reader with information from *The Dictionary of Disasters at Sea During the Age of Steam* by Charles Hocking: "On April 18, 1925, the Japanese steamship *Raifuku Maru* sailed out of Boston, bound for Hamburg with a load of wheat. Shortly after leaving port, the ship encountered a strong storm, and by the morning of April 19 was in dire straits. It sent an SOS signal that was received aboard the White Star Line passenger ship *Homeric*, under Captain Roberts, 70 miles away. Soon afterwards, another radio transmission was received, saying that all lifeboats had been swept off.

"Not long before the *Homeric* saw the ship in distress, a final message was received in broken English: "Dying now, hurry!" Battling huge waves, the liner made its way at a speed of 20 knots to the scene of catastrophe…It was observed that the *Raifuku Maru* was rolling to an angle of 30 degrees, and was completely uncontrollable. Attempting to draw as close as possible, the *Homeric* waited for an opportune moment to save those who managed to remain on the surface of the water; but in such a heavy sea no one succeeded in doing so, and all 48 crew members of the *Raifuku Maru* were drowned."

Kusche doesn't even comment on the event, apparently supposing that a comparison of Hocking's article with the Legend would be to the detriment of the latter. It's hard to believe, however, that the crew of a passenger liner, equipped with all necessary lifesaving accessories, couldn't haul aboard even one of the 48 people who ended up in the water.[42]

---

[42] I will cite another dramatic event at sea, demonstrating how rescue work is carried out in stormy weather. October 9, 1913 in the Atlantic there was a misfortune with the

Considering these events, you come to this conclusion: when the *Homeric* got to the place of the incident, there was not a single live person in the water—only the floating bodies of the dead.

Although Hocking, the compiler of *The Dictionary of Disasters at Sea During the Age of Steam*, doesn't write anything about the "stroke of a dagger," it's doubtful that the *civilizants* pulled these words "out of thin air." The radio transmission from the vessel was in poor English, and during the editing of the source Hocking deleted the words, "This is like the stroke of a dagger!", apparently considering them insignificant.

Meantime, intensive infrasound, as indicated, may elicit unbearable pain in the stomach. It's not for nothing that painful sensations resulting from the perforation of the stomach wall (for example, from an ulcer) are traditionally referred to by surgeons as "daggers."

Shock waves may cause lethal consequences. "Death in these cases results from compression of the brain, disruption of the cardio-vascular system with acute alteration in blood pressure, or the destruction of blood vessels and internal organs."

The examples cited above explain more fully why some "Flying Dutchmen" are found with a full crew, but all dead, without a trace of violence on their bodies.

English steamer "Volturno" with 564 passengers on board. The cargo caught fire in the hold, and soon the fire engulfed the entire ship. Several vessels responded to the SOS signal, among which was the Russian ship Tsar. The Niva magazine (Nr. 43 for 1913) wrote: "Despite the storm…the crew of the Russian steamer set off in boats to the huge fire rushing over the waves and managed to save 102 people. The example of the "Tsar" prompted the teams of other courts to provide more substantial assistance to the victims…523 people were saved. All this happened during a 9-point storm, the height of the waves reached 10 meters!

3.13. "Acoustic Neutrino"

On a December morning in 1944, seven bombers of the American Air Force rose into the air off the east coast of the US, and took a southeast course. The group was heading for Italy to carry out a secret mission. It was still dark, and pilot Dick Stern admired the smooth, silvery expanse of ocean spreading below him. The weather was calm, the engine hummed rhythmically. The other planes flew nearby.

When the squadron was 300 miles out from the Bermudan islands, Dick's plane shook violently. At first he thought he'd hit an air pocket. But now, one after another, a series of blows rained down on the plane, as though an invisible giant was hammering it with his fist. The plane shook, and abruptly lost altitude. Dick was gripped with panic.

"This is the end," flashed through his brain. But then, almost at the water's surface, he managed to regain horizontal flight. Gaining altitude once more, the pilot couldn't see the planes of his comrades. All around stretched the same smooth, silvery water, the same indifferent stars shone in the sky.

Again, Dick was seized by panic, but this time of a mystic nature: the plane was working, the weather was calm, what could that have been? Where did his comrades go? Naturally, Dick didn't fly to Italy, but turned around toward the American shore. At the airbase he learned that, besides himself, only one other plane had returned. The other planes didn't respond to radio contact. The event didn't cause a sensation: it was wartime, and the incident was chalked up to the war situation.

This story has a surprising continuation. In 1961 Stern, now as a passenger, was flying to Miami with his wife. At dinner, just at the moment when he started telling her about the dramatic occurrence he experienced during wartime in that very same region, a similar thing happened. All at once, the plane began to vibrate ominously and lose altitude, but so abruptly that the passengers' dinner ware flew to the ceiling. The vibration continued for nearly a quarter of an hour, and the whole time the plane would gain altitude, then again precipitously fall. Mr. Stern prepared to meet his death: his luck couldn't possibly hold out a second time. However, the plane managed to reach Nassau. Lucky Dick did escape death a second time…

X

I appropriated Dick Stern's aerial adventures from Berlitz's book. The author of bestsellers about the Bermuda Triangle wrapped even these events in a cloak of exceptional mystery, though, judging by his descriptions, what he's talking about is turbulence.

In science, turbulence is defined as "a physical phenomenon pertaining to the difference in pressure at the boundaries of heterogeneous air masses." More simply put, the bumpiness, vibration, and shaking result when the plane intersects air masses of different densities.

Wikipedia states that "turbulence does not, in itself, present any danger to the aircraft," and that "in all the history of aviation, there has not been one instance of a plane crashing due to turbulence."

This is not true; such instances have occurred. The wings of World War II-vintage American B-52 bombers have broken as the result of turbulence. On March 17, 1960, strong shaking resulted in the destruction of the wing of a Northwest Orient Airlines Lockheed L-188C Electra passenger airliner (on a flight from Chicago to Miami). The plane, with 63 people aboard, crashed near the town of Cannelton, Indiana.[43]

Today, extreme durability is factored into the construction of aircraft, and it's thought that the modern airplane can't break apart when it encounters aerial "potholes." But even the sturdiest construction can't prevent falling into an air pothole…

In May of 2017, a Russian passenger airliner was flying from Moscow to Thailand. Shortly before landing, "due to a strong jolt, the aircraft was thrown 200 meters up, after which it fell into an aerial pothole." As a result, 27 passengers received multiple fractures, and were hospitalized in Bangkok…

A more tragic fate befell another Russian airliner. On August 22, 2006, while flying over Donetsk (Ukraine), the Tu-154 was caught in a thunderstorm.

---

[43] According to a report from a US commission on air traffic safety, between 1964 and 1975 there were 183 plane crashes, of which 68 (37%) were connected with turbulence.

At an altitude of 11,860 meters, the plane "was thrown by an ascending current to an extreme altitude, after which it went into a spin. A total of 170 people died."

From the decryption of black boxes: "At 11:32 the airliner experienced strong turbulence with vertical air streams."

"…In 10 seconds, 900 meters of altitude was gained." "The crew had no chance to shift the aircraft to a horizontal flight…"

Flight is sustained by the force of lift—the difference in air pressure over the wing and under it. But now the plane encounters a zone of turbulence, where alongside the disordered vortex motion of the air streams there are strong vertical and horizontal gusts of wind. The pressure on the wing – both above and below—changes abruptly, tossing the aircraft up and down. Moreover, "the nose or tail of the aircraft may ride up, it may be pushed on its side, or it could corkscrew…"

The question arises involuntarily: are infrasonic (shock?) waves generated by seismic activity capable of affecting the hull of an aircraft flying at, say, 10 km altitude?

Obviously, capable. It's not for nothing that physicists called infrasound the "acoustic neutrino," for its ability to penetrate the density of any physical environment, while maintaining its energy for a fairly long time.

In 1962 a direct dependence was discovered between an earthquake that destroyed the Yugoslavian town of Skopje and the condition of the ionosphere: a few hours after each seismic jolt, pronounced radio interference emerged.

Since medium and short radio waves are reflected by the lower layer of the ionosphere (at a height of 50 – 70 kilometers), it was suggested that magnetic disturbances are caused by some terrestrial factor. This factor turned out to be the infrasonic component of the acoustic waves, generated by earthquakes and volcanic eruptions.[44] And as soon as infrasonic waves disturb the ionosphere, they obviously permeate the lower layers of the atmosphere.

---

[44] According to data from the Geo-ecological institute of the Russian Academy of Sciences, 50% of magnetic storms are related to processes occurring in the depths of the Earth. Previously, it was considered that solar flares were the sole cause of magnetic disturbances.

In the summer of 1998, one after another, different regions of the planet experienced earthquakes ranging from 4.0 to 6.5 magnitude. In July alone, 12 air crashes were recorded![45]

By themselves, these coincidences don't prove anything. But they should at least alert the specialists who deal with aviation matters. I have no inkling about aerodynamics, but I do have questions:

How do infrasonic waves from seismic activity influence the physical condition of the atmosphere?

Do they affect the on-board instrument readings?

Wouldn't air masses "inflated" by long-wave lateral fluctuations resemble a thoroughfare paved with inordinately large cobblestones?

Do infrasonic waves add acceleration to vertical gusts of wind in zones of turbulence? And shock waves…

And shock waves? Can a shock wave toss a passenger plane to exorbitant heights?

What do men of science say about these matters? They don't say anything. By admission of the scientists themselves (as of the day of this writing), they "haven't undertaken the study of atmospheric processes with the aid of seismic instruments or barometers"; the effect of infrasonic waves on the readings of such devices "was always perceived as irritating noise."

Three days after the previously mentioned earthquake in Armenia (December 11, 1988), a Russian IL-76 military transport plane carrying humanitarian aid set off for the disaster area. It made its landing approach at the Leninakan airport at night. As it descended, it slammed into Mount Ararat.

---

[45] July 1 – MI-8 helicopter crashes in Buryatia

July 6 – Russian emergency department MI-8 helicopter crashes near Krasnoyarsk

July 13 – MI-8 helicopter crashes in Tatarstan

July 13 – Ukrainian IL-76 plane crashes in the United Arab Emirates

July 18 – IL-78 aircraft crashes during flight from Bulgaria to Eritrea

July 23 – MIG-29 (craft #17) crashes near Budapest, Hungary

July 23 – AN-12 turboprop crashes near St. Petersburg

July 25 – Yak 18-T plane crashes in Korotiche (Ukraine)

July 26 – Russian Yak-54 plane crashes in Alaska

July 28 – Hydroplane crashes in England

July 30 – Beechcraft 19 airliner crashes in France during flight from Lyons to Lorient

July 30 – Passenger helicopter MI-8 crashes near Irkutsk

According to the investigating commission, "the cause of catastrophe was pilot error in setting the altimeter."

That same night, another military transport plane crashed in Armenia – a Yugoslavian AN-12 with a load of medical supplies for victims of the earthquake. During descent, the plane rammed into a highway bridge. Decryption of its black boxes showed that "the height registered on the instruments turned out to have been overestimated by more than 800 meters."

On October 20, 1989, an IL-76TD plane flew out of Ul'yanovsk, bound for Armenia with humanitarian aid cargo. During approach to Leninakan, the crew began its descent in low overcast conditions. "When the plane was at a height of 270 meters (while the altimeter reading showed 1,370 meters), a signal warning of dangerous proximity to the ground was triggered on board…At that moment the crew saw the mountainside straight ahead, but had no time to take emergency measures…"

Reminder to Aerophobes

Why are millions of people afraid of flying? The answer is simple: news of any air catastrophe is immediately taken up by mass news outlets and carried around the world with the aid of hundreds of photos and videos with details of the crash. As a result, audiences and readers get the impression that flying is extraordinarily dangerous. Meantime, the safest mode of transportation is the airplane. Surprising, isn't it? Let's look at the statistics. On average, for every one million flights there is one air disaster, most of which involve small private planes. Therefore, the odds of dying in the crash of a regular commercial flight are vanishingly small. Even flying every day, it would take 21 million years for you to be on that one ill-fated flight that experiences a disaster. At the same time, 1.2 million people die in automobile accidents every year, which is a thousand times more than die in aviation catastrophes. You have a greater chance of ending up in a road accident than dying on a plane.

And apropos of the above. Before takeoff, all pilots receive a weather report for their route. Entering a zone of turbulence is never unexpected. Pilots are ready for it and know the routes to evade or escape it.

3.14. The Quirks of Radio Waves

The connection between the Bermudan enigma and the hypothesis about bioprognosis of earthquakes was something I sensed—I can't put it any other way—when I became acquainted with the books of Berlitz and Kusche. Since then, particular pieces of information, to which I previously paid no attention, began to surface in my memory and, like a jigsaw puzzle, to put themselves together into a logically significant mosaic, illuminating a new path in the Bermudan issue. This path, stretching for thousands of kilometers, connected two oceans—the Atlantic and the Pacific.

The assumption that the course of Flight 19 could have been influenced by some external factor is not groundless. Recall that Lieutenant Wirsching was on a training flight off the Florida coast on that ill-fated day when Taylor's group was searching for salvageable shores. Returning to base, his plane deviated off course by 76 kilometers, for an unknown reason. At the airport it was determined that the navigational instruments of his plane were functioning normally. The assumption of an external factor is confirmed by other events as well.

On December 28, 1948, a DC-3 passenger airliner set out from San Juan for Miami. Following a message from commander Linquist that the plane was within 50 miles of Miami, on-board radio communication was cut off.

According to documents, the DC-3 had been in flight, all the way to its last radio transmission, significantly longer than the estimated time—a highly unusual circumstance, for some reason not noted by Kusche. Meanwhile, the delay in time hints that for some portion of the way the plane had strongly deviated from its assigned route. Judging by the fact that Linquist's last transmission didn't indicate anything out of the ordinary ("everything is fine on board"), it may be concluded that the instrument readings didn't cause any anxiety.

Kusche attempts to explain the fact that the actual course of the DC-3 didn't correspond to onboard instrument readings by citing underdeveloped technology. But the South Korean Boeing KAL-007 was equipped with the latest technology of its time (1983). However, it too went off course for unknown reasons. By 500 kilometers!

Such navigational instruments as onboard radio direction finders and omni-directional beacons on the ground function as the means for reception and emission of radio waves. For this reason, it's logical to suppose that the suggested external factor changes the broadcasting direction of the radio beam that defines the plane's course. This factor may be, for example, a difference in air density.

There's normally a lot of fluctuation of density in the atmosphere. Sunlight is scattered over these non-uniformities, but because of their insignificant measurements, they don't influence the directional distribution of radio waves. For this reason, radio physicists ignore such fluctuations—it's accepted to consider the air density of discrete air masses as sufficiently homogeneous. But sometimes heterogeneities significant in size emerge in the atmosphere…

The route taken by KAL-007 (Anchorage – Seoul) lies over regions of high seismic activity. For example, the Alaskan peninsula experiences nearly five thousand earthquakes per year. From the southern coast of Alaska, a string of earthquake epicenters stretches along the bottom of the Bering Sea—along the Aleutian island chain—to Kamchatka.[46] Kamchatka is joined to Japan by a gigantic mountain ridge lying along the bottom of the Pacific Ocean. And deep underwater trenches lie along the Kuril Islands. Such topography of the ocean floor is characteristic of subduction zones, in which active volcanoes and earthquake epicenters are concentrated.[47]

Earthquakes are usually preceded and followed by a series of underground jolts. The latter "give a whole" concert with changing rhythm, frequency, and strength of the impact; they may last several days, weeks, even months.» It's

---

[46] The Aleutian Islands comprise an archipelago of volcanic origin, forming a "chain" from the coast of Alaska to the base of the Kamchatka peninsula. The majority of strong earthquakes in the region occur along the Aleutian chain, which is one of the most seismically active structures on Earth.

[47] A subduction zone is an extended zone along which one gigantic block of the Earth's crust is drawn beneath another.

easy to imagine that lateral infrasonic waves generated by convulsions of the ocean floor intersect with jet streams.

Jet streams are swift torrents of air at heights from 7 to 16 kilometers, that can circulate in the troposphere for thousands of kilometers. The breadth of these "aerial rivers" reaches 1,000 kilometers in places, while the speed of the wind in them varies from 150 to 900 kilometers/hour. Pilots often use these irregular jet streams to save fuel and flight time.

Jet streams occur most often at the border between the mainland and ocean, where there's a sharp differential in the temperature of air masses. A significant portion of the route taken by the Korean Boeing KAL-007 follows exactly along the coastlines of the western Pacific. For most of the way the airliner is flying on autopilot, at a height of 10 – 12 kilometers. Its course is maintained by navigational instruments, as mentioned before, through the reception and emission of radio waves.

…And so, infrasonic waves reach a height at which jet streams prevail. Here, caught up by the powerful aerial torrent, they will now spread along the earth's surface. In this manner, a series of compressed and uncompacted air patches will form in the jet stream. In a corridor like this, a radio beam will deviate more and more from a linear path.

A visual demonstration of the above may be served by an optical effect often described in school physics textbooks: a teaspoon, lowered into a glass of water, looks distorted to the eye. This illusion is caused by the breaking of light rays at the border between two environments—air and water. Radio waves, possessing the same physical nature as light, also break at the border of different densities.

Moreover, the longer the wave, the more pronounced the break. This is why, in an air mass in which infrasonic waves are spreading, extremely significant durations of compression and rarefaction will prevail. Under these conditions, the radio beam determining the course of an airplane will behave just like the beam of light streaming from the spoon in the glass of water: at the boundaries of different air densities it will change the direction of its transmission. And the more such aerial compressions and rarefactions along the route of a plane, the more the radio beam will deviate…

Light beams are perceived by the eye, and their breaking may be told by the effect of the "crooked spoon." Bur a person can't see radio waves. They are directly received by navigational instruments. But these cold instruments "don't notice" that the running wave has changed direction. How can the pilot know that his instruments are deceiving him? He doesn't know—and the plane deviates more and more from its course.

Some physicist may observe that the described effect of the "crooked spoon" is too simple, too sketchy to serve as an explanation for an event comprising so many riddles. Well, reality almost always turns out to be more complicated than any diagram. Still, we won't shrink from the attempt to "fit" our version to other peculiar circumstances of the incident of August 31, 1983.

"The zone of invisibility"

The fact that KAL-007 was able to penetrate deep into Soviet airspace undetected received various explanations. Thus, the journal *Counterspy* wrote that Soviet radar was subjected to deliberate jamming; this must be why the "spy" plane went unobserved for so long.[48] Other evidence indicates that several radar installations on Kamchatka malfunctioned that day. Generally speaking, this may not be excluded.

But, after all, the Korean airliner's violation of Soviet airspace was likewise "overlooked" by both American and Japanese radar. And later two aircraft (KAL-007 and Osipovich's fighter jet) disappeared for a time from radar screens on Sakhalin. Such coincidences indicate that the cause, apparently, is not in the technology…

Lieutenant-colonel Osipovich spoke of a "zone of invisibility" without going into the physical nature of this phenomenon. However, an object in the

---

[48] Deliberate jamming has been carried out since the end of the Second World War, with the goal of disrupting the functioning of the enemy's radar. It's produced by two types of radio jammers – emission of impulses on the same frequency as the enemy's radar station, or emission of chaotic, "noisy" signals. If, on the day of the incident, the air defense service on Kamchatka had registered signals on the same frequency as their monitoring stations, then the premeditation of the jamming would have been indisputable, and would have been a decisive argument from Moscow. However, the Soviet side said nothing of this. Which means that the jamming of radar on Kamchatka (if it took place) could have been carried out only by "noisy" signals. But distinguishing static caused by external factors from "noise" generated by the malfunctioning of the radio apparatus itself is impossible without special devices. Was this why the enraged Moscow leadership found convenient scapegoats in the senior officers of the Far East air defense? Some of them were accused of sloppiness (malfunctions in Army technology!) and demoted in rank.

air will remain unnoticed if infrasonic waves are running between it and the radar. Meanwhile, radar screens "light up" with horizontal lines, bright rings, or other bizarre configurations.

Finally, though, the intruder is discovered. Ground services launch interceptor fighter jets at the target within a range of hundreds of meters. The pilot of the fighter jet, having attained the designated height, can pinpoint the location of the target, even in conditions of poor visibility, with the help of onboard radar. So why, on the day of the incident, did everything on the Soviet border "go haywire"?

Radar intercepts objects found within the horizon line. But under heterogeneous atmospheric conditions, even short waves may spread over distances exceeding the line of sight. An analogous phenomenon is familiar in optics: when light beams break at the boundary between warm and cold layers of air, objects found far from the line of sight show up in the field of vision.

An incident that occurred near the end of the Second World War is both interesting and instructive.

"The Americans were preparing to seize the island of Kiska, occupied by the Japanese, in the Pacific Ocean. The American fleet was located within 600 miles of the island when radar operators suddenly discovered a mysterious squadron. It was located at a distance of only 40 – 50 miles. At the battle alarm signal, the ships' crews prepared to mount a defense. But soon the squadron vanished.

"A few days later, the American fleet reached Kiska. No enemy was in sight. The mysterious squadron, it turned out, had been evacuating the Japanese garrison. Due to a radio mirage, the operators were mistaken by 550 miles. Had they known about the quirks of radio wave transmission, it's probable that the American fleet could have conducted a successful operation."

Within the framework of our mental experiment we may put forth the following proposition: on August 31, 1983, the silhouette of an intruder appeared on the radar screens of the Soviet air defense, but in actuality the airliner was already far beyond the horizon line—only its radio mirage still lurked on the screens. Isn't this why the ground services sent their fighter jets to a region of the sky where "the KAL-007 wasn't, and couldn't be"?

"The zone of radio silence"

In the sky over Sakhalin, the KAL-007 was in a zone of "radio silence": the Korean pilots didn't hear Osipovich's radio transmissions. They didn't hear, apparently, for the same reason the Soviet and American radar "went blind." Generally speaking, the band of frequencies used by pilots in emergency situations is hardly ever subject to interference. But according to the confession of Osipovich himself, made by him during the period of Gorbachev's glasnost, he didn't switch his radio transmitter to the emergency frequency—"because of a lack of time."

X

The physicist, of course, is right: the mental experiment with the "crooked spoon" is only a diagram. But here is a circumstance that, I suspect, may have a direct relationship to the event under discussion: on August 17, 1983, an earthquake of magnitude 6.8 occurred on Kamchatka.

3.15. Tellurian Currents

October 5, 1960 was the day when World War Three almost began. A group of unidentified objects appeared on the radar screens of the American military base at Thule, Greenland. The UFOs were moving swiftly to the west, in the direction of the United States of America.

Ufologist John Keel, author of the book *Operation Trojan Horse*, describes the dramatic events: "The red telephones at Strategic Aviation Headquarters in Omaha, Nebraska shrilly rang out, and bomber crews on air bases all over the world scrambled to their planes. With nuclear weapons on board, the B-52 bombers took off with a roar, and ominously circled their air bases, awaiting final orders that would identify targets for a responding attack on Soviet territory. Strategic Aviation Headquarters persistently called Thule for confirmation of received information. There was no answer from the base. Could it be that Thule had already been attacked?…"

Fortunately, atomic war was averted: the UFOs suddenly vanished from radar screens, and the severed contact between continents was explained by an iceberg having cut the underwater cable. "It's extremely strange that an iceberg, never before having been the cause of an accident, picked this particular time for its 'sabotage,'" remarks Keel, hinting with transparent irony at the machinations of wily extraterrestrials.

Indeed, the coincidence in timing allows one to think that there must be a connection between the appearance of a group of UFOs and the damage to the cable. But not the one suggested by the ufologist…

On April 25, 1966, a few hours before a catastrophic earthquake in Tashkent (Uzbekistan), an "electrical storm" broke out: overhead wires sparked, the surfaces of apartment walls glowed, fluorescent lamps spontaneously came on. Instruments at the hydrometeorological station began to trace crooked lines similar to those that appear before thunderstorms. The sky above the city also lit up. The aurora above the city was a whitish-pink color, reminiscent of light scattered by lightning. And during the earthquake itself, many residents, running in panic between ruined houses, were witness to a group of UFOs—glowing spheres, floating in the sky like hot-air balloons.

X

American meteorologist William Jackson Humphreys once enumerated various phenomena that could be taken for ball lightning. There turned out to be 11 of them.

"So, does this mean there isn't anything that may be called ball lightning?"

His answer: "I don't know. I only know that many things may be called ball lightning, that actually turn out to be something else."

Indeed, the glowing objects formed with the aid of natural electricity exhibit a wide diversity of dimension, length of duration, and speed of movement. For example, "classical" ball lightning seldom exceeds the diameter of a soccer ball, and "lives" a comparatively brief life of a few minutes, after which it usually explodes. At the same time, there are numerous reports of "long-lived" glowing objects, such as plasmoids, whose diameter can exceed several meters.

As pointed out in 2007 by Alexander Tatarinov, chief researcher of the Geological Institute of the Siberian Branch of the Russian Academy of Sciences, "plasmoids manifest...along fractures of the Earth, generated by seismic activity. They consist of ionized gases, charged dust particles, and aerosols. These glowing formations burst from the Earth's depths with great speed." [49]

Yevgeny Barkovsky, staff scientist at the Institute of Earth Physics of the Russian Academy of Sciences shares the same opinion: "Below the Earth's surface, a powerful local discharge of static electricity formed... Simultaneously an electromagnetic formation may arise in the atmosphere, visually suggesting an UFO. So we cannot exclude the possibility that the ball lightning and "saucers," at the core, may be artifacts of electrophysical

---

[49] Some of Tatarinov's colleagues were skeptical toward his hypothesis. However, it was later upheld by geophysicists from a number of leading Russian and foreign research centers: Russian Academy of Sciences Institute of Earth Physics, Moscow physical engineering institute, Russian Federation Ministry of Defense, Cambridge University (United Kingdom), and others.

processes, similar in kind, but differing in power—one arising in the atmosphere, the other in the Earth's crust.

"Bursting forth out of the depths of the Earth, a plasmoid takes the most stable form available, becoming a sphere. In size it may range from five centimeters to several meters in diameter."

Plasmoids have a peculiar nucleus and a less dense breadth, noticeably stretched out along the line of Earth's magnetic field—a veritable "flying saucer"! And, by the way, it flies: the energy field surrounding plasmoids generates its own unique magnetic pillow, on which they slide along lines of geomagnetic force. Like invisible rails, these magnetic pillows allow plasmoids to move in air and water environments with incredible speeds, as well as exhibiting marvelous maneuverability that seems to defy all laws of inertia. These glowing spheres are often what eyewitnesses identify as "UFOs."

By all indications, the majority of UFOs consist of optical manifestations of electrical processes, occupying enormous expanses in the atmosphere. Therefore, if the spectrum of our sense of sight suddenly became wider, we would see plasmoids in the form of gigantic spheres, whose glow would get dimmer according to its distance from the center.

X

The great ancient philosopher Plato instructed us: if Earth were to be viewed from the cosmos, it would resemble a soccer ball, sewn together out of several scraps of leather. As it turns out, the surface of our planet is indeed covered with dozens of blocks of the Earth's crust, that is, tectonic plates, divided from each other by deep geological fractures.[50] (We can only guess from what sources the ancients got evidence about the construction of the Earth, corroborated by modern science.)

In breadth, tectonic plates can be as large as 10,000 kilometers, while their depth varies from 60 to 100 kilometers. These gigantic blocks of the Earth's crust—oceanic or continental—are in constant motion. They come together, rub against each other, or slide, one underneath the other. Increasing seismic tension along the edges of the plates is periodically being "eased" by barely

---

[50] A lithospheric (or tectonic) plate is a massive chunk of the Earth's crust, part of the solid covering of the Earth
(lithosphere), bordered by zones of seismic, volcanic, and tectonic activity.

felt jolts, or by tectonic and volcanic quakes. Evidently, the generation of terrestrial (telluric) currents also takes place here.

It is believed that earth currents are generated under the influence of the so-called piezoelectric effect in quartz compounds, which are abundant—up to 60 percent—in the Earth's crust. The piezoelectric effect consists in the fact that when quartz plates are deformed ("piezo" in Greek means "to press"), electrical charges of opposite sign appear on their faces[51]. In short, the value of electric forces increases in those areas where there are movements (friction, squeezing) of the Earth's crust.

X

Four months after the Tashkent earthquake, workers at the local seismic station decided to conduct an observation of the aftershocks that continued to issue periodically from the depths of the Earth[52]. For this, they drilled a half-kilometer borehole on the grounds of the station, and lowered a seismometer on the end of a cable protected from external disturbances. After a few days, they noticed intense leakage of electrical charge from the external loose ends of the cable core.

"It turned out that the central cable wire was short-circuiting on the braided shield, half a kilometer down…The reason for the disruption was the manifestation of a large electrical potential difference between the braided shield and the central cable wire."

Apparently, in accordance with the same "electrical scheme," the underwater cable was damaged on October 5, 1960, near the coast of Greenland. And it was no coincidence that the cable connection was disrupted at the same time that a group of UFOs appeared over Greenland.

Analogous damage to a cable connection likely contributed its share to the tragedy that occurred in the Bermuda Triangle on December 5, 1945. Let's recall: when the disoriented planes of Flight 19 were finally located, their coordinates should have been immediately reported to search headquarters.

---

[51] The piezoelectric effect was discovered by Pierre Curie in 1880

[52] The repeated shock following the main seismic shock are called aftershocks by geophysicists. Accordingly, an earthquake is preceded by foreshocks. The number and intensity of both aftershocks and foreshocks decrease over time, but they can last for months.

But just at that "critical period," the cable connection between the air and sea services failed.

From the scientific literature:

«Tellurian currents create extreme tension between landing points of telephone-telegraph cables, sometimes leading to the breaking of connections, or even damage to the apparatus».

3.16. Lightning Bolts from the Underworld

Toward the end of the 1960s, world media were overflowing with reports of unidentified submerged objects (USOs), registered in different regions of the planet. Moreover, it wasn't only journalists inclined to sensationalism who were telling about them, but submarine officers and representatives of intelligence services.

Rear Admiral U. Beketov, former K-19 nuclear submarine commander: "…We emerged close to the US and moved toward the Bermudan Islands…We encountered many unusual phenomena: sometimes the instruments went out of whack, sometimes there were powerful disturbances. Some unexplained phenomena were apparently artificially generated. More than once, our instruments registered material objects moving with unimaginable speed…in the neighborhood of 400 kilometers per hour. As though these objects weren't subject to the laws of physics…"

Captain 1st rank N. Tushin: "I myself, like many submarine commanders, saw glowing rings in the ocean…I'm certain that our K-219 boat could have been destroyed by the impact of an unidentified submerged object…"

In the summer of 1961, a Soviet K-19 atomic submarine was conducting training exercises in the Norwegian Sea. On board the sub were three ballistic missiles armed with nuclear warheads.

The nightmare began on the 16th day of the sailing, at 4 hours, 7 minutes on the 4th of July. Because of water leakage in the reactor's cooling system, the fuel rods began to rapidly heat up, threatening an imminent nuclear explosion. Since at that time the boat was located within 70 miles of the NATO base on the Norwegian island of Jan Mayen, such an explosion may have been taken for a nuclear attack. This was a time of military standoff between the USSR and the USA, the very peak of the cold war, and—no doubt about it—the response would have been immediate.

They decided on the only correct solution—to doctor the cooling system with materials at hand. Eight volunteers entered the nuclear compartment, where the level of radiation was lethal to humans, and in 15 – 20 minutes had repaired the malfunction. At the cost of their lives, the Soviet sailors prevented a global nuclear catastrophe.[53]

From the conclusion of the commission of the USSR Ministry of Defense: "…The ship went into service with a crack in its pipeline that originated in the factory. During the course of operation…the crack widened due to agitation, until the line [in the cooling system of the reactor] burst."

The uninformed reader may take such an explanation on faith. But for someone who knows how scrupulously technical control was observed at defense plants, this conclusion ("factory defect") raises doubts.

As it happens, an analogous conclusion was reached by the commission looking into the loss of the American nuclear submarine *Thresher*, that drowned in the Atlantic on April 10, 1963. On the basis of ultrasound flaw detection and underwater photographs of the submarine's remains, the commission determined the probable cause of the catastrophe: "due to a crack in a welding seam in the seawater pipelines…there was a sharp drop of energy in the pump conveying cold water to the reactor."

In the autumn of 1986, a Soviet K-219 nuclear submarine with nuclear ballistic missiles on board, was on military patrol in the Sargasso Sea. On the 3rd of October, a fire broke out in the missile silo. Anticipating the poisoning of his crew by toxic vapors, commander Igor Britanov ordered the boat to the surface.

"A deep extraneous groove was observed on the caps of the missile silos, ending in the cap of the emergency shaft, where the rocket had caught fire."

---

[53] A day after the emergency, the K-19 crew was evacuated to destroyers that arrived on the scene. Eight people who had averted a meltdown died of radiation poisoning. All other crew members had received a high dose of radiation. In the interests of secrecy, the official diagnosis was altered, and the sailors signed a nondisclosure agreement.

Moreover, the furrow had a bluish tint, rather than the metallic color that results from a collision with a solid object.

A few days after the fire on the K-219, the atomic submarine *Augusta* was put into dry dock for repairs at a US Air Force base; it had sustained damage, as was reported, "upon collision with an undetermined object." A conjecture arose that this "object" had been the Soviet K-219. However, in 1986, when another Soviet submarine, a K-457, experienced an incident in the Greenland Sea, an extraneous trace identical to the one found on the K-219, was discovered on the caps of its missile silos, running from stern to bow.

Just such a trace was discovered in May of 1974 on the caps of the missile silos of a K-408 Soviet nuclear submarine that had collided, as is supposed, with the American nuclear submarine *Pintado* near the coast of Kamchatka.

The main purpose of submarines—in case of a sudden onset of war— is to destroy the enemy submarine with torpedoes before it has a chance to inflict a strike from underwater. For this reason, submarines stalk each other, often at dangerously close distances. Dozens of submarines belonging to various countries have been damaged, or have drowned as a result of undetermined causes. But if an enemy sub is found to be in the vicinity of the one that has suffered a catastrophe, the investigating committees naturally adopt what to them seems the more likely scenario—a collision.

X

In 2013, materials pertaining to an investigation into the disappearance of the Israeli submarine *Dakar* near the island of Crete were declassified. They included a curious report from the Navy command of Great Britain. The report says, in part: "On January 26, the crew of a fishing vessel near Crete observed a strange submerged object in the vicinity where the *Dakar* perished. It was large and shone brightly, but moved with absolute silence." According to the conclusion of the investigative commission, the submarine *Dakar* evidently perished after colliding with an unidentified submerged object.

Two days after the disappearance of the *Dakar*, the French Navy was conducting exercises in the Mediterranean. At one point the submarine *Minerva* reported that it had begun to follow an unknown object that had appeared on the sonar screen. At a depth of 60 meters, communication with the submarine broke off. Dozens of ships and planes from five countries took part in the search, but not the slightest trace of the *Minerva* was found. An unidentified submerged object continuously appeared and promptly disappeared in the search area.

A special group for the study of mysterious phenomena in the Earth's oceans was established in the investigative administration of the USSR Navy. Military analysts constantly reported to their superiors: USOs are top secret NATO operations, designed for secret sabotage. Judging by the bluish traces on the bodies of submarines, the sabotage was carried out by targeted jets of plasma (similar traces are usually left by a plasma torch, cutting through metal at great speed). There was, in fact, another version put forth: USOs are plasma formations of natural origin, akin to ball lightning.

Indeed, ball lightning often leaves traces on gas and water pipelines. At least one incident is described when ball lightning burned a round opening in the side of a factory pipe.

In 1965, an LI-2 military transport plane, flying over the Kola Peninsula, encountered ball lightning. "…The magnetic compass needle tossed chaotically from side to side for 3 – 5 minutes. When it finally stopped, its readings turned out to be incorrect. The entire radio apparatus also went out of order. Examination of the airplane after landing revealed…two openings melted into the back edge of the horizontal stabilizer.

"In Ulyanovsk, in July of 1989, ball lightning 'attacked' a streetcar…At the driver's command, the passengers were able to get out in time, and a few seconds later, under its belly…ball lightning flashed. There was a sound of rending metal, and the body of the streetcar began to fall apart as though under a blowtorch."

One can imagine the surprise of passengers: ball lightning occurs in the atmosphere—this is understandable. But for fireballs to "attack" the tram…

3.17. Mysterious Explosions

On August 4, 2020, at approximately six p.m., a terrifying explosion in the port district of Beirut demolished the entire coastal district and its infrastructure. The blast wave caused major damage to numerous buildings in the city. More than two hundred people were killed and over six thousand injured. According to the Lebanese media, ammonium nitrate stored in a warehouse at the port had caught fire.

"Security officials have repeatedly warned the president and prime minister that thousands of tons of explosives could blow up and destruct half of Beirut…" The port administration was accused of criminal negligence and the Lebanese government resigned.

"Israeli geophysicists came closer to the truth behind the events in Beirut. Analyzing seismic readings from transmitters located 70 km from Beirut, in the depths of the sea, not far from Cyprus, they arrived at a sensational conclusion: six mysterious explosions had preceded the massive detonation at the port. Approximate equal in force, they went off, one after the other, in eleven-second intervals. Then, forty-three seconds later, came the catastrophic blast that destroyed a large portion of the Beirut port. The foreign geophysicists, having checked and rechecked the data obtained in Israel, confirmed that the large-scale explosion had, indeed, been preceded by six smaller blasts, "and no doubt about this remains"."

The Israelis also calculated that "the explosion that led to this level of destruction was far beyond the limits of any simple explosion of chemical substances." And there is also the fact that 2700 tons of ammonium nitrate at the epicenter of an explosion are incapable of creating a crater forty-three meters deep.

Thus, all this casts doubt upon the official version that the explosion of ammonium nitrate was caused by the fire in the port.

X

Five years before the Beirut catastrophe, a similar event transpired in Tianjin, China, a city of fifteen million people. On the evening of August 12, 2015, in the port, where tons of toxic substances were stored in warehouses, two explosions occurred, about thirty-second apart. The second was far more powerful, sending a mushroom cloud billowing into the sky that was registered by equipment on a space station in orbit around Earth. The blast wave annihilated residential districts a kilometer from the port. On a lot close to the warehouses, ten thousand automobiles were incinerated.

The entire territory within a four-kilometer radius of the epicenter was scorched and contaminated. One hundred and fourteen people were killed and over seven hundred injured. The authorities admitted that hundreds of tons of sodium cyanide, an extremely hazardous chemical substance, had been stored at the Tianjin port. The first responders, arriving to aid the victims, were unaware that they were working in a contaminated zone. The consequences were so horrifying that reporters dubbed the event "Little Chernobyl."

According to investigators, the cause of the disaster was a fire that arose spontaneously when a container overheated in the summer sun, upon which the flames spread to the warehouses containing the toxic substances. However, there is a vulnerability in their conclusion: sodium cyanide does not explode when subjected to fire, it merely emits toxic fumes. The Minister of Public Security of the People's Republic of China, Guo Shenkong, had no choice but to explain the cause of the disaster as a neglect of safety precautions, and promise to "severely punish" those responsible...

Meanwhile, officials from the China Seismological Center and the US Geological Survey's monitoring department in Beijing conducted a thorough study of the huge, "apparently bottomless" crater at the epicenter of the explosion. According to experts, the depth of the funnel and the cracks formed inside it "could have arisen only in the event of an underground explosion with an epicenter at a shallow depth." Apparently, the explosions in Tianjin were caused by seismic activity. It is noteworthy that a deep funnel formed at the epicenter of the explosion in Beirut...

According to Greenpeace, only in the first half of 2015 in China there were at least 14 serious accidents and explosions at chemical plants and coal mines.

Who can guarantee that these catastrophes, which have taken place over the course of several months, are due only to criminal negligence?

X

On April 20, 2001, a Russian television station ran a documentary program entitled "Chernobyl: the Doomed Nuclear Power Plant." The film made a profound impression upon viewers—it turned out that the reason for the disaster was a super weak earthquake. The evidence presented was irrefutable: seismogram readings taken on the day of the catastrophe. The seismograms attested to the fact that on April 26, 1986, in the Chernobyl region, there was an earthquake registering 1.6 on the Richter scale. Typical of such "quiet" (or "creeping") earthquakes is a slight rise in the ground level, together with scattered depressions.

This does not tend to lead to destruction on the surface; in fact, such events generally pass almost unnoticed. But if a large technological facility is located in the area, the report went, even an extremely weak earthquake can precipitate an accident. To be completely accurate, the catastrophe was in fact caused not by the earthquake itself, but by electrical energy released from deep within the earth.[54]

Inside the chamber, underneath the reactor, were gaping holes two meters across in the concrete floor and walls. Judging by the melted edges of the holes, the concrete had dissolved (which can happen at temperatures of at least several thousand degrees), while metal pipes and supports attached to the reactor had simply evaporated. At the same time many non-metallic objects remained unharmed.

Such an outcome is typical of a ball lightning impact. It would seem that the nuclear reactor was "strafed" from beneath, from underground. And that is not all. Several hours before the explosion, a "flying saucer" was seen over the

---

[54] In accordance with standards developed in the USSR, technological facilities are not built closer than fifty kilometers from a geological fault. The Chernobyl Nuclear Power Plant, however, had been raised in proximity to the juncture of several deep geological faults belonging to the Pripyat system. Seismologists have repeatedly applied to the relevant authorities, pointed out the potential for an accident, and strongly recommended mothballing the station. The officials, however, ignored these recommendations.

nuclear power plant. It hung in the air for six hours and was observed by hundreds.

Igor Yanitsky, a leading geophysicist at the Russian Ministry of Defense: "…on the night of April 25, seismic activity in the area of the Chernobyl nuclear power plant increased precipitously…The fault propelled a plasmoid up into the pedestal of the reactor, causing approximately twenty highly powerful discharges, accompanied by blindingly bright bursts of light and explosions."

Possessing enormous charges, plasmoids are capable of disrupting any technical construction. The operation of a power unit on the surface of the earth or in sea water can greatly increase the probability of emissions of electromagnetic energy from the bowels of the Earth.

As with ball lightning, plasmoids "are capable of penetrating barriers and entering closed spaces. Having their own magnetic field, they can affect technology even at a distance. When a plasmoid appears close to airplanes and ships, fuses blow, batteries and accumulators are discharged, and radio communications are interrupted…fires or explosions may occur on board."

X

On August 10, 2000, in the Barents Sea, the Russian navy began a series of maneuvers. One submarine, the Kursk, entered the sea to execute a series of training shots at an imaginary enemy. On August 12, the Kursk sank. A search brigade located the ship the following day. In the upper part of the hull gaped a "perfectly round aperture," and the nose of the vessel had been destroyed. All one hundred and eighteen crew members died. According to the experts' conclusions, the catastrophe was the result of an explosion. At 11:28 am, Moscow time, a 6576-A training torpedo had detonated in the nose section of the Kursk.[55]

---

[55] A number of authoritative persons opposed the conclusion of the commission of inquiry. Admiral Eduard Baltin, former commander in chief of the Black Sea fleet, for instance, asserts that "a torpedo explosion aboard the nuclear submarine was not the proximate cause of the catastrophe that followed. Over the entire forty-year history of

Cruising at a depth of thirty meters, the submarine plummeted to the bottom, a depth of one hundred and eight meters. There, a second explosion occurred, destroying the front half of the submarine…

The seismograms at the Norwegian NORSAR seismographic station showed, however, that in the region where the Kursk was located, there were, in actuality…three explosions. Consequently, one of them occurred not inside, but outside the ship, close by the hull.

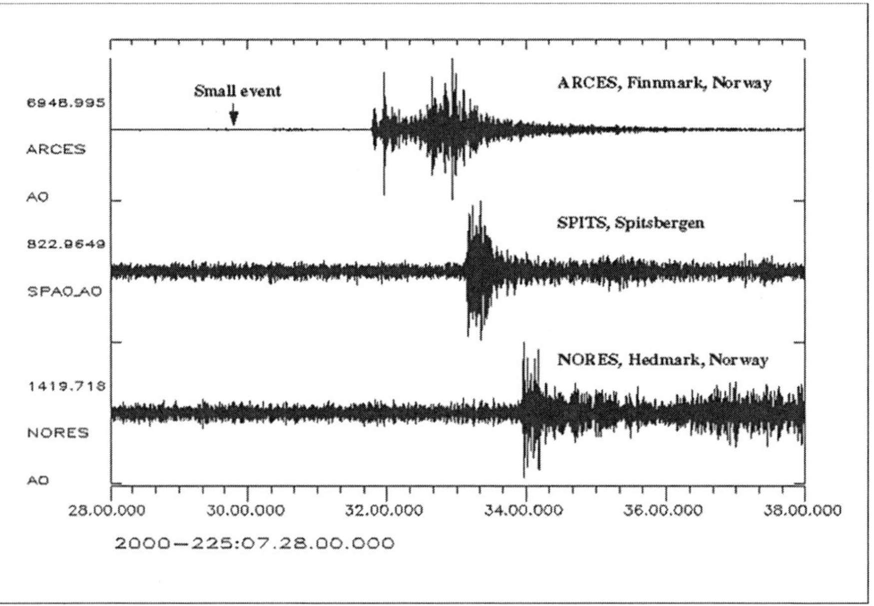

Between May and August 2000, in the basin of the northern seas, active tectonic processes were registered. It is noteworthy that shortly before the explosions on board the Kursk submarine, an earthquake with an intensity of 3.8 on the Richter scale occurred in the Barents Sea, off the coast of Norway. And seismic activity, as mentioned, causes the generation of telluric currents, hence the formation of "electric balls". Clearly, the Kursk plummeted to the

---

the use of these torpedoes, not one has ever exploded, and they are not capable of exploding on their own."

Professor Sergei Proshkin, senior submarine weaponry designer, declared: "The explosion of the fuel components of a 65-76 torpedo, as a result of which the nuclear submarine Kursk perished, could have happened only as the result of external action directed at torpedo."

bottom in close proximity to a tectonic fault which held a "ripening" plasmoid with a powerful energy potential…

The fact that the cause of the destruction of the Kursk could be a plasmoid is evidenced by the observations of Soviet submariners during the Great Patriotic War: "Submarine crew members frequently and systematically reported small spheres of ball lighting arising within the closed space of a submarine. They would appear either when electric batteries were activated, or when a high-inductance electrical motors were deactivated or improperly engaged. Attempts to reproduce this phenomenon using the submarine's reserve battery either failed or led to an explosion."

PS. In the light of the above facts, another version suggests itself about the cause of the damage to the Nord Stream gas pipelines, where on September 26, 2022, seismographs registered two "shocks" in the Baltic Sea.

3.18. You Shouldn't Believe Your Eyes...

We can't, of course, suspect all eyewitnesses of UFOs of fraud or mental illness, just as we can't suspect all informants to the courts of the Inquisition of ill-will toward their neighbors. But we can suppose that many of those who took moonlit smoke from a cottage chimney to be a witch on a broomstick, or UFOs to be extraterrestrial space ships, were genuinely deluded. Such delusion has connections to peculiarities in the human psyche, and so, as a psychiatrist, I'll attempt to illuminate certain mechanisms that give rise to them as clearly as possible.

At the beginning of this book, I allowed myself to compare the Bermudan enigma to the mystery of love. To the extent that love is an eternal mystery, it has been interpreted in various ways by artists of all times.

"Love is a subjective emotion," wrote Stendhal, "and it depends in large part on the lover, more than the beloved…In Salzburg, they leave a dry branch stuck in the salt pits. It's an ugly black branch, but when approached on the following day, it's completely covered with salt crystals. And the whole branch sparkles, it gladdens and excites the gaze…"

Our beloved is like that branch. By herself, the woman we love may not be anything special, but the power of love decorates the object of our passion with sparkling crystals, and she appears to us completely differently than she is in reality.

A graphic illustration of the subjectivity of love may be found in the passion inflaming Don Quixote toward Dulcinea del Toboso. The Spanish hidalgo, never having even met his chosen one, endowed her with the qualities which, in his opinion, noble ladies should possess. Meantime, Dulcinea had none of these qualities, being an ordinary provincial.

Another French writer, André Maurois, develops Stendhal's line of thought: "Love is a form of insanity, because when we fall in love with a woman we no longer see her as she is in reality." Although this is only a poetic interpretation of the subjective perception of the emotion of love, it's in complete accord with the scientific interpretation of known psychopathological phenomena characterized by a distorted perception of the real object. The term "illusion" is frequently employed in colloquial speech.

"He's creating illusions," for example, is what might be said about an infatuated youth when he endows the girl next door with unrealistic merits. But in psychiatry, illusions are understood precisely as tricks of perception, usually visual in character. Illusions may be observed even in psychically healthy people. It's easy to imagine a child in a poorly lighted entry way perceiving his father's coat on the hanger to be a lurking stranger.

Illusions (like hallucinations) in many ways represent a blind spot in psychopathology. "In the majority of cases of false perception, we don't know the causes giving rise to them, we know nothing of the conditions under which they manifest," the famous German psychiatrist, Karl Jaspers, wrote in 1913. Today, however, it's possible to try to clarify a few things.

The manifestation of illusions is often connected to a strong passion, or disturbance of the emotional sphere. Indeed, Don Quixote suffered from a maniacal state. Besides, he was gripped by delusional ideas: he saw his foes in everything. This is why Cervantes' hero entered into his famous combat with the windmills.

And so, besides the emotional factor, a substantial role in tricks of perception is played by corresponding representations; representations arising in the mind are precisely what conditions the meaningful content of illusions. I'll clarify this with our example of the child. Let's say that, from an overheard conversation between his parents, the boy learned that the neighboring apartment had been robbed by burglars. In this case, the boy is more likely to "see" the silhouette of a lurking burglar in the hanging coat, than anything else.[56]

---

[56] Sometimes, dominating representations serve as points of departure for attempts to explain a different puzzling phenomenon. Is this valid? One may argue: "Not always, of course; however, if there are points of contact…" But points of contact, if diligently sought after, will always be found. Besides, a person can always accidentally hit the "bullseye," and then his idea escalates into a serious natural science concept (the legend of Newton's apple). But this approach to a problem, when it's forced to the desired conclusion, more often results in mistakes of judgment and incorrect

A third prerequisite for the manifestation of illusion is a deficit of information. In the case of the child, a deficit of information may have been the consequence of poor lighting, due to which some details of the hanging coat weren't recognized. If the child has an active imagination, and there was an added element of fear (passion!), then the unrecognized details will immediately be filled in with representations of burglars. This is the same way illusionary perceptions may occur in people when they notice an unfamiliar object in the sky.

From the narrative of a witness of a UFO: "…On the 27th of April, my family and I were returning home. Suddenly my wife and I saw a bright-white, glowing object in the sky, approaching us from the northwest. Its speed was so great that, before my wife could say "It's a flying saucer," it was descending to a minimum elevation. Soon it was possible to see lighted windows in it…And here I understood that it was necessary for me to have other witnesses of this event. I called the police and asked them to raise an alarm at the airports located in that direction; they took my request into consideration.

"I got in my car and tried to follow after this mysterious object…It occurred to me that this incident would make an interesting newspaper article. I called the police again and asked whether appropriate authorities had been informed of what had taken place. I was told that they were thinking about it…I then told them that I'd call the airport at Selfridge myself, which I did. As you see, I got the unenviable role of the person everyone wants to get rid of…"

If an observer recognized a flying object as, say, a plane, he would think nothing of it. But the object isn't recognized. Here is where the deficit of information comes into play, and consequently a readiness to supply that deficit with corresponding representations. This last has long been formulated in the heads of millions of people.

Over the decades, so much has been said and written about UFOs in connection with cosmic visitors, that many people have somehow ceased to be

---

deductions. Even Charles Berlitz couldn't avoid this mistake, when he "saw" a manmade pyramid in the photograph of a conical protrusion on the bottom of the Bermuda Triangle. Gripped by a passion to see traces of an unknown civilization in everything, he didn't think that the photograph may have captured a (likewise conical) natural object such as a volcano, of which there are many on the floor of the Northern Atlantic.

conscious of what the three letters stand for. But they're the initials of words that, in a scientific sense, construct an extremely correct definition: unidentified flying objects. **Unidentified!** But for many people, unidentified flying objects turn into **identified** space ships.

And finally—the emotional piece. An encounter with a Mystery! The individual has heard and read so much about UFOs, and here, unexpectedly, he himself is meeting face to face with this extraordinary phenomenon. There's the UFO, flying, flying in his direction…In anticipation of even more unusual events, the eyewitness is seized with a rapture bordering on exultation. In that moment the observer is conscious…yes, yes!– of his own personal accountability to humanity. He's obliged to report to society everything he is now seeing, and what may possibly come of it. He has no right to omit the smallest detail; everything may turn out to be of the utmost importance…

And now, in the flickering of the object as it flies through the sky, the observer "distinguishes" the blinking of signal lights, and in its uneven brightness he "sees" illuminated portholes and the glinting of a metallic body. These impressions grow like a snowball, incorporating fragments of visual sensations and their interpretations, mixed with corresponding recollections and the products of fantasy.

This fantasizing may be unintentional when the person sincerely believes in the reality of his perceptions. In such a case, under the influence of a disproportionately inflamed imagination, he may even take a group of people in the distance to be passengers of UFOs. But now the emotional surge has died down, and the witness begins to comprehend that what he's seen is actually not enough: he won't be able to convince anyone that what he saw was really a space ship.

But he himself is absolutely sure of it! Then the eyewitness begins to augment his "absolutely authentic" observations with what are now fabricated details. Thus are born thousands of testimonies regarding space ships and encounters with extraterrestrials.

"Well, what about it?" asks the disappointed reader. "Does the author himself deny the existence of UFOs? So, the extraterrestrials and their technology are all fabrications and delusions?"

I can't answer these questions unequivocally; after all, this book isn't focused on the issue of UFOs. I can only assert that the nature of Bermudan phenomena may be explained without involving testimonies from ufology.

"But what about the unsolved mysteries?" the reader persists. "Why, for example, when American pilots fired on UFOs, did their planes explode? Does this not indicate that UFOs are controlled by intelligent beings? And why does the pulsating light emitted by a UFO cause people to feel fear and panic? And how to explain the fact that there were no pilots or passengers in the South Korean Boeing that was shot down by Osipovich?"

Let's go step by step. As mentioned, plasma formations have a significant electrical potential. A missile launched in the direction of a UFO ionizes the air, thereby creating an electrically conductive corridor. Through this corridor, the energy of the plasmoid is discharged. Such a boomerang effect may well destroy the "aggressor" aircraft.

As for the pulsating light coming from the UFO…It is noteworthy that with the same frequency—2 hertz—longitudinal seismic waves propagate, which geophysicists call P-waves[57]. Possessing high energy, they cover the entire Earth, penetrating through the water column and air. Apparently, P-waves (through the piezoelectric effect) "impose" their frequencies on telluric currents, as well as on plasma formations in the atmosphere and aquatic environment.

"And what about the South Korean "Boeing"?" reminds the reader. "Suitcases – found, clothes – found, but there were no bodies of the dead…

"Regarding the Korean "Boeing"…Here I must confess: it was the author's trick, so to speak, a literary technique. I cited only those newspaper articles, the authors of which hurried to report the absence of the bodies of the dead. Later publications said that during the search for traces of the catastrophe, human corpses were found in sea water. I emphasize: I did not invent anything from myself, I only cited official sources!"

A terribly offensive confession for many lovers of secrets. But after all, I meant it for the best, I wanted to make the book more interesting, and I did my best for the reader. You can, of course, argue about the legitimacy, correctness, of such a technique. However, I ask the reader to be indulgent to me—after all, Charles Berlitz and Lawrence Kusche used exactly the same technique (and to a much greater extent!) to substantiate their versions.

And also the author of an article reporting on the discovery of Flight 19 planes under water – there were no bodies of pilots under the glass caps. Of

---

[57] Due to the high speed of propagation, P-waves got their name from the Latin "prima" – primary.

course, this is amazing. But the mystery quickly dissipates if we take into account that the Avengers were not examined, they were only photographed at a depth of 250 meters. 45 years have passed since the day of the disaster, during which time not only the clothes and bodies of people decayed in the sea grave, but also their skeletons crumbled.

It will be especially interesting for the reader to know that the sources of infrasound are the processes occurring not only in the earth's crust, but also in the airspace: auroras, geomagnetic field disturbances, pulsating currents in the ionosphere and, of course, "UFO"-plasmoids.

Part IV
The Bermuda Triangle
Changes its Address

In recent years, publications about the Triangle have become noticeably rarer, apparently because nothing unusual has happened there for a long time. Meanwhile, there are increased reports of earthquakes, volcanic eruptions, and technological catastrophes in Southeast Asia. These two conditions allow an examination of the Bermudan enigma, so to speak, from another side.

According to publications, the peak of seismic activity in Southeast Asia occurred between 2011 and 2015. Curiosity impelled me to "take a stroll" through the Internet. It turns out: between February of 2011 and August of 2015, in the waters and on the islands of the Malaysian archipelago, there were 13 aviation catastrophes! These weren't some sort of little planes belonging to lovers of aeronautics. By no means. We're talking about scheduled flights, primarily passenger, operated by professional pilots.[58]

---

[58] **2011.**

February 12. An Indonesian Airlines Aviocar-100 passenger airliner crashed.

May 7. An Indonesian passenger airliner Xian-MA60 fell into the sea and sank.

September 9. An Indonesian Susi Air Cessna-208 cargo plane was destroyed when it hit a mountain.

Catastrophes involving modern air vessels are rare. But when, in the course of four and a half years, 11 passenger and 2 military planes go down within a restricted area!...The theory of probability hints: all these air catastrophes probably have a common cause.

To test the hypothesis, I turned to the cases of the death of two passenger aircraft. These tragic events drew the special attention of analysts in different countries. At the same time, the media included enough factual material for me to make use of.

September 29. An Indonesian Aviocar-200 passenger airliner crashed.

October 13. A PNG Air Dash 8 Q100 passenger airliner crashed on the northern coast of Papua New Guinea.

2012.

May 9. A Russian Super Jet-100 passenger airliner crashed during a demonstration flight near Jakarta, Indonesia.

June 21. An Indonesian military Fokker-F27 plane went down in a residential area of Jakarta.

2014.

March 8. A Malaysian Airlines Boeing 777 passenger jetliner, heading for Beijing, disappeared over the Indian Ocean.

July 23. A TransAsia Airways ATR-72 passenger airliner plowed into the ground in Taiwan.

December 28. A Malaysian AirAsia Airbus-320, heading out of Surabaya, Indonesia for Singapore, "fell into a tailspin," and went down near the coast of Sumatra. (That same day there was an incident with another AirAsia passenger airliner. "The Airbus-320 was making a flight along an internal route in Malaysia, but returned soon after takeoff 'due to technical problems.'")

2015.

February 4. A TransAsia passenger airliner fell into the Keelung River, flowing through the Taiwanese capital of Taipei, due to engine failure.

June 30. An Indonesian Hercules C-130 military transport plane crashed into a residential area of the city Medan, in northern Sumatra.

August 16. An Indonesian Trigana Air passenger airliner crashed into Mount Tangok in eastern Indonesia.

4.1. The Captive of Mount Salak

A Russian Sukhoi Super Jet-100, making a demonstration flight on March 9, 2012, disappeared from radar screens 21 minutes after takeoff from the airport at Jakarta. Its fragments were found on a steep slope of Mount Salak, some 60 kilometers south of the Indonesian capital. 45 people perished.

A year before the tragedy, a model Super Jet-100 had been exhibited at the prestigious air show in Farnsborough, near London. The air show set a record for the number of participants and concluded deals. Civilian Aircraft of Sukhoi signed a sales contract with an Indonesian aviation company for the delivery of Super Jet-100s, totaling billions of dollars. Thanks to this deal, the company was counting on cornering the markets of Southeast Asia. The disaster dealt a blow to the reputation of the Russian aviation company, founded in 2006 at the initiative of President Putin.

Most often, the causes of air tragedies lie in human factors and technological malfunctions. This is why the commission investigating the catastrophe turned first of all to analyzing these "traditional" theories...

Theory one (the human factor)

Russian vice-premier, Dmitry Rogozin, overseer of the military-industrial complex, hastened to put forward the human factor theory. This was understandable: a version like that excludes technical malfunctions, and was thus advantageous to the Russian side.

"Many experts suggest that the Super Jet-100 encountered a thunderstorm in the mountains. The planes commander, apparently, decided not to raise the plane, but chose to avoid the cloud from below. He asked permission to

descend and (most important) received it." [59] Less than a minute later, communication with the aircraft broke off.

An array of Russian newspapers points to "pilot error": *"the maneuver of descending and making a sharp right turn at a height lower than the mountain peaks greatly increased the danger of collision—it was only a few seconds of flight to Mount Salak."*

One may argue here. First, there's no proof that radio contact broke off at the moment of catastrophe (radio contact is often disrupted in areas of thunderstorms). Second, for an experienced pilot, even a few seconds are enough to avoid colliding with a mountain. And the crew members, demonstrating the achievements of their country's aviation industry, were doubtless among the highest professionals.

So, who made the mistake—the Russian pilot, or the Indonesian air traffic controller who gave the "go ahead" for the descent? *"There's no consensus, as yet, whose mistake could have led to the catastrophe."*

Theory two (technical malfunctions)

On May 14, one of the "black boxes" was recovered. On the basis of recorded conversations between the crew members, the Russo-Indonesian investigating committee concluded: *"the aircraft was fully operational until the immediate collision with the mountain. The terrain warning system TAWS was engaged and functional. An answer to the question, why a highly experienced Russian crew failed to react to the audio signal warning of danger…so far has not been found."*

If the audio warning system was engaged and functional right up to the collision, then the involuntary thought arises: whether all the crew members suddenly went deaf, or were all unconscious…

And here's an intriguing circumstance: *"the second tape, which records the functionality of the plane's equipment, and is the one considered by experts to be the most important, the Russian side now declines to continue searching for."*

The Russian hero, test pilot Anatoly Knyshev: "That they didn't find the parametric onboard tape recorder, just to have it on hand for the designers…If

---

[59] Citations from Russian media are given in cursive, here and below.

something went wrong with the technology, they'll try to hide that fact to the last."

"The Indonesians, in their turn, also don't want to be responsible for killing the hopes of the Russian aviation industry, so they've announced that they'll continue searching for the tape that recorded the parameters of the airliner's flight path."

Later, the second tape recorder was found. It didn't contain any information indicating technical difficulties.

Theory three (technical sabotage)

The catastrophe of the Russian passenger airliner in Indonesia during demonstration flights put the experts in a stalemate. Some commentators have labeled it "unexplainable." And unexplainable events, especially if they involve big money and politics, always generate conspiracy theories…

On May 24 the newspaper *Komsomol'skaya Pravda* reported: *"Russian military intelligence is investigating whether the Americans could have provoked the catastrophe of the Russian Sukhoi airliner."*

The fact is, there's an American Air Force base at the Jakarta airport, from which the Sukhoi jet took off. From this geographical fact, Russian power brokers spun a simple logical chain.

"Who stands to gain from a catastrophe with the Super Jet-100?"

"Obviously, the competitor of the Russian aviation industry, the American company Boeing."

"So maybe it was the Americans who engineered a prank?"

"Sure, why not, it's totally possible."

"The chief investigation department of the foreign intelligence service reports that it's possible to effect instrument failure in a plane from the ground. Moreover, intelligence offices are sure that such an 'attack' can be proved. That's what they're working on now."

An anonymous general from Russian Military Intelligence: *"We've long been following the work of the US Air Force in Jakarta. We know that they have special technical developments that can disrupt a connection or cause a malfunction in the settings on a plane from the ground; for instance, an airliner is flying at a specific altitude, but after interference from the ground the instruments show something different. It's possible this couldn't have happened without such 'interference.'"*

This sabotage theory assumes that the Indonesian flight controller who gave the "go ahead" for descent was participating in the scheme. It also reveals the unprecedented cynicism of the American military.

"Over the past few months, some among the Russian media have been fanning the flames of anti-American sentiment against the background of Putin's return to the Kremlin. Suspicions toward the US have their roots in the cold war of the Soviet era."

Theory four (essentially, the author's)

The newspaper, *Jakarta Post*, called Mount Salak "a graveyard of planes"—over a period of ten years (2001 – 2012) it was the site of seven aviation catastrophes. Analysts attribute the loss of airplanes to high turbulence and swiftly changing weather conditions in the mountain area. However, such an explanation doesn't stand up to scrutiny: during the same period, no other mountain in other regions of the Earth registered such frequent air disasters…

Could it be that the problem lies not in weather conditions, but in the geophysical peculiarities of the region? Let's check it out.

The Malaysian archipelago, the largest in the world and including islands such as Java, Sumatra, New Guinea, and others—is a segment of the so-called "Pacific Ring of Fire." Bordering the Pacific Ocean, this is a chain of active volcanoes and earthquake centers stretching some 40,000 kilometers. The ocean floor is streaked with long-drawn tectonic fractures, along which lie nearly 90% of the planet's volcanoes and earthquakes.

Mount Salak, with which the Russian plane collided, is an active volcano in the western part of the island of Java. 12 kilometers to the north of the volcano is the city of Bogor, which has suffered more than once from Salak's activity, and from forest fires occasioned by eruptions…

The informed reader, guessing where the author is going with this, will note: on May 9, the volcano didn't erupt. Yes, that was so. However, let's not be hasty.

Like tectonic fractures, the mouths of volcanoes, as well as fissures in their slopes, act as wave generators through which electromagnetic energy of the

Earth's interior escapes into the atmosphere. Moreover, the same seismic (ionized) clouds, in which onboard instruments malfunction or fail completely, form in the vicinity of volcanoes. So, this is where the "thundercloud" on the Super Jet-100's flight path came from!

In general, atmospheric electricity presents a serious danger to aviation. Lightning striking a plane may cause it to crash, or its fuel tanks to ignite. For this reason, the decision of the Super Jet-100 pilot to bypass the suspicious cloud from below makes sense. But in circumventing the visible threat, the pilot didn't realize that the real danger for him may be lurking in the invisible sections of the electromagnetic field.

It's doubtful that the Indonesian flight controller was in cahoots with American intelligence services. It seems that the locator at the Jakarta airport didn't track the plane's proximity to the mountain. Being convinced that the plane was higher than it actually was, the flight controller lightly permitted the commander his fateful maneuver. He was judging not by actual data, but by what he'd received from the last radar signal.

It seems the locator on board the Super Jet was also "blinded"; this is why the audio system connected to it failed to signal impending danger. And when visual contact with the mountain was actually made, it was too late…

And how can we forget the aviation catastrophe that occurred four days after the Spitak earthquake! A Russian IL-76 military transport plane, making a landing, slammed into Mount Ararat—the tallest volcanic massif of the Armenian plateau.[60]

---

[60] In mountainous regions, bedrock accumulates a large reserve of elastic energy, which elevates the piezoelectric effect. Magnetic disturbances in electric fields above mountainous regions corrupt the readings of onboard instruments. Because of this, planes collide with mountains relatively often. Thus, in 1996, while coming in for a landing on Spitsbergen Island, a TU-154 airliner slammed into a mountain. The same thing happened in 1997 to a Yak-42 in Greece. In 2001, a Yak-40 collided with a mountain in Iran, and in 2002, also in the mountains of Iran, an AN-140 and a TU-154 suffered catastrophes. All indicated areas of aviation catastrophes comprise zones of heightened seismic activity.

4.2. Where Satan Runs the Show

On March 8, 2014, a Malaysian Air Boeing 777 airliner (Flight MN-370), with 239 people aboard, rose into the air from the airport at Kuala Lumpur at 00:41 local time, heading for Beijing. The airliner never arrived at its destination.

The search for the Malaysian Boeing 777, involving the planes and ships of 26 countries, was the most expensive in aviation history. Over 100 square miles of ocean surface were searched. Later, the search territory was broadened to include the waters of the Indian Ocean. And—no results.

"The disappearance of the Malaysian Boeing shook the whole world and stupefied the specialists of many countries," the newspapers wrote. "It's hard to believe, but in our time, with such tremendous technical potential at the place of arrival, the fate of 227 passengers and 12 crew members of the ill-fated flight remain a mystery."

Let's try to measure our seismotectonic hypothesis against this event. To begin with, we'll establish its chronology.

…About an hour after takeoff, when the airliner was over the South China Sea, ground services registered three dramatic facts:

- Radio contact with the Boeing was cut off.
- Transponder signals, conveying information about the location of the plane and its identification data, ceased.
- The Boeing disappeared from radar screens.

Since these events took place more or less simultaneously, dispatchers at tracking stations suggested that the airliner had suffered a catastrophe over the South China Sea. Soon, however, it was picked up by Malaysian military radar. For an unknown reason, the Boeing made a left turn and proceeded basically the way it had come. It crossed Malaysia and entered the Malacca Strait, dividing the western coast of Malaysia and the island of Sumatra.

Over the Malacca Strait, the airliner rose to an altitude of 15 kilometers (some data say 12 kilometers), after which it descended to an altitude of 3.6 kilometers, and disappeared from radar screens once more. Trackers on the ground then concluded there were terrorists aboard, aiming to hijack the passenger Boeing; they disconnected the communication systems and sharply reduced altitude in order to escape being detected by radar.

However, a question arises here: why did the airliner vanish from radar screens when it was flying at an altitude of 10.5 kilometers over the South China Sea? And further. Let's allow that the alleged hijackers disconnected onboard communication systems. But they couldn't have affected the working of ground radar! Impossible. Consequently, the radio apparatus of the Boeing, as well as ground radar, experienced the effect of some external factor. It's noteworthy that the tragedy was preceded by "artillery preparation."

January 4, 2014. The Sinabung Volcano erupts on the island of Sumatra.

February 14. A powerful eruption of the Kelud volcano takes place on the island of Java. Hundreds of thousands of inhabitants within a 10-kilometer radius of the volcano's crater are evacuated.

February 16. The Anak-Krakatau volcano becomes active on the northwest coast of Java.

February 25. An earthquake occurs near the coast of the Indonesian island of Ternate…

"A minimum of 19 other volcanoes in the country have also expressed a desire to begin erupting."

What the analysts didn't teach.

It's surprising that analysts paid no attention to this fact: on the night of March 8, when the Malaysian Boeing was over the Malacca Strait, an earthquake of 5.5 magnitude was registered in the Java island region.

It's true that anomalies with the airliner already arose when it was over the South China Sea, some 500 kilometers from Sumatra. However, the earthquake itself was only a local tectonic manifestation. In reality, tectonic dislocations, as a rule, embrace wide expanses of the Earth's crust. On the day after the airliner's disappearance, earthquakes were registered in the Kuril Islands, and a little later—in Japan and China. Conclusion: magnetic disturbances and electrical fields sporadically arose in the atmosphere over various sections of the western Pacific and Southeast Asia.

One other circumstance: over the South China Sea, in the region from where the last message from the Boeing passenger liner was transmitted, Thailand radar registered a UFO.

Apparently, a dangerous situation developed on board. This could have taken the form of disruptions in the control system, fire, or an explosion, resulting in depressurization. Some reason caused the pilots to turn their airliner in the opposite direction, to land at the nearest airport…

Where to search?

Some readers, following the news, will argue: nearest airport?! But later signals from the Boeing were registered in the southern Indian Ocean, some seven thousand kilometers from its last disappearance. The airliner's flight path south was also shown by satellite data. Don't these facts confirm the hijacking theory?

To my view, they don't. If the airliner were hijacked, they wouldn't have blown it up in midair. To think that the hijackers landed it somewhere also doesn't compute—an enormous airliner with 239 people aboard, wherever it landed, would have been impossible to conceal for long.

Concerning the radio signals, this factor also contains some ambiguities. Search vessels registered suspicious signals in different parts of the Indian Ocean, so that experts couldn't unequivocally confirm their connection to the vanished plane. They also had a hard time explaining how a Boeing 777 could fly unnoticed over several thousand kilometers containing a great quantity of military ships with their radar systems.

Regarding the satellite data. This was purely technical information about the condition of the plane's engines, not including its location. On the basis of this information, analysts determined two vectors, indicating merely the possible location of the airliner at the moment it crashed. One vector went south, over the Indian Ocean; the other went north, crossing Thailand, Vietnam, China, and Kazakhstan. Since the governments of these countries indicated that the airliner hadn't entered their airspace, it was presumed that the plane flew to the southern portion of the Indian Ocean.

Relatives of some passengers tried to get through to them, and heard beeps in the receiver. However, none of those flying on this ill-fated flight answered. I am inclined to the opinion of those experts who believe that there were no survivors on board the Boeing during the last hours of the flight.

"Analysts choose their words extremely carefully in discussing the possible hypotheses of this air catastrophe, citing lack of information…The reasons why search and rescue crews from 26 countries have thus far not been able to discover [the Malaysian Boeing's] fragments are obvious. The waters comprising this search include parts of the Indian and Pacific Oceans, and also several seas. Considering the circumstance that…ocean currents may have carried the fragments for tens and hundreds of miles from the scene of catastrophe, we're forced to assert: finding traces of the crash will be more difficult than finding a needle in a haystack."[61]

---

[61] Two years after the tragedy in the Indian Ocean, two floating objects were found, which were identified as fragments of the vanished airliner.

4.3. Do Prophecies Come True?

Among the prophecies of Leonardo da Vinci may be found the following: "The waters of the ocean will rise to the mountaintops…The waters will cause the destruction of cities…many of the greatest nations will be drowned in their own homes."

And here's what was said by the earlier mentioned American clairvoyant, Edgar Cayce: "If it seems to you that the something is moving and rumbling in the bowels of the Earth, that is the beginning of the displacement of the Earth's interior core." Back in 1936, Cayce beheld the destruction of humanity in the year 2120, as the result of powerful earthquakes that will entirely destroy all the continents.

According to the prophecies of other soothsayers, the Earth will begin to convulse at the start of the 21st century. "Earthquakes, lava, and tsunamis will overrun the face of our planet; its climate and magnetic poles will change."

The data may differ, but all forecasters agree on one thing: the 21st century will become a time of harshest trials for humanity. The people will enter a stretch of prolonged economic crises, rampaging natural elements, global climate change, and military confrontations.

One may take the clairvoyants in a variety of ways. Let's see, however, whether their prophecies come true…

Situated inside the Earth is an iron-nickel core with a radius of nearly 3.5 thousand kilometers. Like the yolk inside an egg, the core "floats" in a partially melted mantle.[62] The spinning of the core is responsible for the creation of the planet's magnetic field. In 2008, coworkers at the Helmholtz Geophysical Research Institute (Germany), as well as the Cosmic Dawn Center in Denmark,

---

[62] The mantle is part of the geosphere, lying between the crust and core of the planet. It contains the major part of Earth's substances.

registered precipitous dislocation of the core toward the Pacific Ocean. Moreover, the dislocation occurred too rapidly, and it continues to accelerate.

Such a redistribution of gigantic masses causes changes in the magnetic field of the planet. So, since the end of the 20th century, the magnetic field intensity between Australia and Antarctica has increased, and on the opposite side of the Earth, in the western Atlantic, on the contrary, it has decreased by 10%. Is it not for this reason that in the 21st century a "tectonic calm" established in the Bermuda Triangle, but the frequency and intensity of earthquakes in China, Japan, and Indonesia increased?

December 26, 2004

A huge earthquake, one of the top three most powerful ever recorded in the history of observations, occurred near the western coast of Sumatra. The tsunami it generated reached the coasts of Sri Lanka, India, Thailand, and even the United Arab Republics, located some 7 thousand kilometers from the epicenter. This tsunami is considered one of the deadliest in recent history, killing more than 300 thousand people in 12 countries situated around the Indian Ocean. The exact toll of the dead will probably never be known, since a majority were swept into the ocean.

March 11, 2011

A magnitude 9 earthquake occurred in the Pacific Ocean near the eastern shore of the Japanese island of Honshu. This most powerful earthquake in the history of the Land of the Rising Sun received the official name, "Great Japanese Earthquake." Roughly an hour after the first jolts, a tsunami 13 meters high overwhelmed the nuclear power station Fukushima-1, with fatal consequences—the reactor's emergency cooling system failed, after which there was a series of radioactive explosions.

The disaster at Fukushima-1 was a major radiation catastrophe, registering at the maximum level of 7 on the International scale of atomic events. Over 17 thousand people were killed, or disappeared without a trace. According to satellite data, the earthquake displaced the northeast portion of the island of Honshu by 5.3 meters to the southeast.

April 11, 2012

In the region of Sumatra…"the Earth cracked." This apocalyptic event was reported by all the world's news agencies. As written in the journal *Nature*, "A series of extremely powerful jolts in the vicinity of the Indonesian island of Sumatra in April of 2012 turned out to be the 'footprints' of tectonic processes that split the Indo-Australian lithospheric plate into 'Asian' and 'Australian' portions."

Specialists noted that the Sumatran earthquake was the most powerful in its "weight category"—enormous masses of the Earth's crust were displaced by 40 meters in relation to each other. The US Geological Survey predicted that the consequences of the splitting of this tectonic plate would be felt in other corners of the Earth.

Literally a few days after this event, it was as though a pointer had run around the perimeter of the Pacific "ring of fire": from mid-April until June, there were earthquakes in Thailand, Japan, Sakhalin, Peru, Chile, and New Guinea.

August 12, 2015

The most massive technological catastrophe in the history of China occurs in the megalopolis of Tianjin.

Here it is appropriate to also recall, that from February of 2011 to August of 2015, 13 aviation catastrophes took place in the waters of the Malaysian archipelago. All of them, apparently, represent links in the chain of abovementioned "seismic events."

4.4. "Sore Spot" of America

In recent years, the Pacific lithospheric plate has begun to sink faster under the North American plate, on which the United States and Canada are located. At the same time, the colossal seismic stress at the boundaries of the plates brings a terrible catastrophe closer—the eruption of the Yellowstone volcano.

Over the past years, the Pacific lithospheric plate has accelerated its subduction under the North American plate, on which the United States and Canada are located. At the same time, the colossal seismic stress at the boundaries of the plates is hastening the approach of a frightful calamity—the eruption of the Yellowstone volcano. Its caldera is located under Yellowstone National Park, in the state of Wyoming. Unlike ordinary volcanoes, it occupies an enormous territory, with an area of 30 by 100 kilometers.

The entire valley floor is filled with incandescent magma under a thinning layer of Earth's crust. Since 2004 the volcano, which hasn't erupted in over 600,000 years, has begun to show signs of activity. Underground jolts have increased in frequency. Over the past few years, the soil has acquired a 180-centimeter bulge due to pressure from magma. Today, the magma is so close to the surface that areas of very hot soil, deep cracks, and hot geysers have begun to appear in the park. Some areas of the park have been closed to visitors, having been deemed life-threatening.

Since the spring of 2014, bison have begun to abandon the national park, followed by deer. This is a sure sign of impending catastrophe: unlike people, animals unerringly sense natural cataclysms.

Under the Administration of the President of the United States, a scientific council was created, involving leading volcanologists and seismologists. Monitoring was carried out under the National Security Council, the Department of Defense, and representatives from the intelligence services, endowed with full authority. The scientific council gives monthly reports to the US President regarding the situation in Yellowstone. According to the opinion of NASA scientists, the eruption of a super volcano is much more dangerous for humanity than the threat of a comet fall. The explosion will be more powerful than one hundred Hiroshimas.

In the first minutes after the start of the eruption, about 200 thousand people will die. But these are very insignificant losses in comparison with the number of victims of the raging forces of nature. The explosion of the Yellowstone super volcano will provoke the eruption of hundreds of ordinary volcanoes and tsunamis around the planet, which will carry off millions of human lives.

A huge amount of ash tossed into the atmosphere will block the sunlight, the world will plunge into darkness. In two or three weeks after the eruption, the average temperature on the planet's surface will sink to -30 degrees C. It is hard to even imagine what it will be like for the surviving people in the conditions of the onset nuclear winter. And the most frightening thing is that the explosion of a super volcano can occur any minute…

4.5. Oh, GOD!

I hesitated for a long time whether to remove this last chapter from the book or leave it. The doubt is reasonable: the book is devoted to a natural-science problem, and in conclusion—prophecies, prayers, and mystical coincidences. There is cognitive dissonance. Nevertheless, I decided to leave the chapter, having preceded it with a brief explanation.

I am fully aware that I risk "tarnishing" my reputation in the eyes of friends and acquaintances engaged in the field of science. Nevertheless, with all doubts, the desire to be heard by other people also outweighed people who read the Holy Scriptures, who put God's commandments above the notorious "sanity", believing that the "Higher Mind" rules the World.

Many of these people are pessimistic about the future, seeing in it the inevitability of the Apocalypse, which we seem to be accelerating. Moreover, I want to give them hope for a better future...

I can easily imagine how other representatives of the natural sciences, reading these lines, raise their eyebrows in surprise. To these men, I would remind the well-known metaphor of the physicist Karl Heisenberg, Nobel laureate: "the first sip from the vessel of natural sciences makes us atheists, but at the bottom of the vessel, God awaits us." It is helpful to add here: that in the face of death, even a convinced atheist turns to God...

X

Our ancestors intuitively felt that the Earth is a living organism. Trying to live in harmony with nature, they did not kill totem animals and did not cut down sacred groves. In the twentieth century, when technology rapidly developed, a proud man declared himself the king of nature. Only a small handful of "advanced" people know that our planet is a highly developed, self-organizing system sensitive to ongoing events. They have long noticed: where there is social tension, mutual hatred of people, where military conflicts arise, earthquakes, volcanic eruptions, hurricanes, floods occur more often…

The Earth has endured violence against itself for a long time. It suffered tests of atomic and geophysical weapons, drilling of deep wells, emptying underground oil lakes, predatory deforestation, and burying radioactive and toxic substances in the ocean depths.

However, recently, when people's aggression approached the line beyond which the shadow of an atomic war loomed, the planet began to lose patience. Increasing natural disasters are taking the lives of thousands and thousands of people. Waking up from a centuries-old sleep, the Yolluston volcano is the last menacing warning to humankind!

Are we all going to die! Will we perish already in the current generation?! It is scary to even think about it. Nevertheless, if you think about it…

1281. Genghis Khan's grandson Kublai Khan decided to conquer Japan. When the vast Mongol fleet approached the Land of the Rising Sun, the Japanese priests in their temples began to pray for salvation. Moreover, a miracle happened. An unexpectedly flown typhoon hit the Mongolian ships and destroyed them.

1541. In Spain, the Inquisition is atrocious. Under the pain of death, the Jews of this country are forced to accept baptism. Those suspected of secretly performing Jewish rituals are burned at stake. The Spanish armada of hundreds of ships with thousands of troops is approaching the shores of Algiers. On October 23, the Jews of Algeria, realizing what awaited them, gathered in synagogues and prayed to God for salvation. A few hours later, the sky turned black, and a terrible storm broke out on the sea. Almost all Spanish ships sank or crashed on the coastal rocks.

1944. When the US 3rd Fleet was approaching the coast of Japan, the Cabinet of Ministers of that country called on all the people to perform a prayer ceremony. The appeal stated that the simultaneous prayer of 100 million Japanese could avert the threat of invasion.

The command of the American Navy rolled with laughter when they learned about such an exotic defensive act of the enemy. Nevertheless, soon the American sailors were no longer laughing. An unprecedented typhoon hit their armored armada, sinking three destroyers and damaging 28 ships. Hurricane winds blew 146 aircraft off the decks of aircraft carriers. The operation was aborted.

1991. Iraqi dictator Saddam Hussein invaded Kuwait, which he considered his province. On January 16, 1991, after the expiration of the UN ultimatum, the aviation of the international coalition from the USA and NATO countries began massive bombardments of Iraq. In response, Saddam Hussein decided to…punish Israel. The next day, January 17, several Scud-type ballistic missiles were fired at Tel Aviv and Haifa. The United States persuaded the Jewish state not to respond to provocations so that the event would not take on the character of an Arab-Israeli clash. During those dramatic days, millions of Israelis prayed to God for their lives.

In total, Iraq fired 39 (!) ballistic missiles at Israel. Result: One person died of a heart attack during an air raid alert.

Perhaps the above examples are just coincidences. What if it is not random? What if the mental energy of a large number of people can really influence the course of events?…

Somehow, doctors at the University of Durham (North Carolina, USA) undertook to study the effect of prayer on the human body. They invited several nurses, monks, and priests of various faiths to pray for 700 heart patients. For several days, doctors kept diaries about the condition of these patients. After the experiment was completed, the results of medical observations were handed over to the experts. "In 500 patients, prayers increased the recovery rate by about 93%."

Someone will express doubts about the purity of the experiment. They say the patients could know that they were being prayed for. In this case, it cannot be ruled out that the improvement in their condition was due to a psychological factor – faith in the magical effectiveness of prayer. Well, the doubt is reasonable. Nevertheless, how then evaluate prayer's effect on pathogenic bacteria, not only in the human body but also in the aquatic environment?

For several years, members of the Union "Orthodox Scientists of Russia" have been studying the properties of such phenomena as the power of the sign of the cross, the word of God, and the healing properties of holy water. For the study, water samples were taken from various reservoirs: wells, rivers, and lakes, which contained Escherichia coli and Staphylococcus aureus. After reading the appropriate prayers, "the number of harmful bacteria decreases by almost a hundred times…"

I do not presume to judge how objectively the researchers evaluated the results of experiments that run counter to official science. However, who will undertake to assert that such a thing is impossible in principle?

Had it been my will, I would have called an emergency meeting of the UN Security Council. From a high rostrum, I would call on the governments of all countries to address their citizens through newspapers, radio, television, and computer networks. So that all earthlings – Christians, Jews, Muslims, Buddhists—stop strife and appeal to the Almighty. Moreover, the Merciful Lord, perhaps, will hear the prayer of billions of his unreasonable children: – God, save our planet! Oh, God…

Bibliography

Berlitz Ch. Das Bermuda-Dreieck, Wien/Hamburg, 1974.

Berlitz Ch. Spurlos, Wien/Hamburg, 1977.

Group D. Beweise: das Bermuda-Dreieck. München, 1987.

Winer R. Vom teufels Dreieck zum teufels Rachen. München, 1978.

Baulin I. Behind the Barrier of Hearing (in Russian). Moscow, 1971.

Belousov V. Earth, its structure and development (in Russian). Moscow, 1963.

Brianchaninov I. About wonders and signs (in Russian). Yaroslavl, 1870.

Washington I. The life and voyages of Christopher Columbus (in Russian). Kharkov, 1992.

Wiener N. Cybernetics and society (in Russian). Moscow, 1958.

Gangnus A. The mystery of earthly disasters (in Russian). Moscow, 1977.

Giliarovsky V. The doctrine of hallucinations (in Russian). Moscow, 1949.

Golitsyn G., Chugunov E. Acoustic-gravity waves in the atmosphere. The USSR Academy of Sciences. Institute of Atmospheric Physics (in Russian). Moscow, 1973.

Drake Ch. Et al. The ocean itself and for us (in Russian). Moscow, 1982.

Ermolenko S.I. Natural vibrations of the atmosphere and the Earth according to barometric and seismometric data (Thesis) (in Russian). Saint Petersburg, 2016.

Zelig K. Albert Einstein (in Russian). Moscow, 1966.

Isakovich M.A., Shmakova N.E. Infrasound. Literature review for the period 1968 – 1977. (in Russian) Moscow, 1978.

Karpenko M. The universe is intelligent (in Russian). Moscow, 1992.

Kiel D. UFO. The campaign "Trojan Horse" (in Russian). Saint Petersburg, 1992.

Kuznetsov V.V. On the connection between earthquakes and atmospheric electricity (in Russian). Vestnik KRAUNTS. Phyisical and mathematical sciences, 2017, No. 2(18), 99-110.

Kuzovkin A. Unidentified objects (in Russian). Moscow, 1990.

Kuzovkin A., Nepomnyashchy N. What happened to the destroyer Eldridge (in Russian). Moscow, 1991.

Cousteau J., Dumas F. The silent world (in Russian). Moscow, 1957.

Kusche L. The Bermuda Triangle Mystery—Solved (in Russian). Moscow, 1978.

Carrington R. A Biography of the Sea (in Russian). Saint Petersburg, 1966.

Linkov E.M. et al. Seismo-gravitational oscillations of the Earth and associated atmospheric perturbations (in Russian). Doklady AN SSSR, 1991, vol. 306, pp. 314-317.

Malina Ya., Malinova R. Great Mysteries of the Earth (in Russian). Moscow, 1993.

Medvedev S., Shchebalin N. You can argue with an earthquake (in Russian). Moscow, 1967.

Myasnikov L. Inaudible sound (in Russian). Saint Petersburg, 1967.

Novogrudsky et al. Infrasound: enemy or friend (in Russian). Moscow, 1989.

Obruchev V. Fundamentals of geology (in Russian). Moscow, 1947.

Ostrovsky B. Phenomenon-psi-2 (in Russian). Newspaper "Vecherny Baku", 22.12.1964.

Ostrovsky B. The voice of silence (in Russian). Journal "Znanie – sila", № 7, 1969.

Ostrovsky B. On the effect of infrasonic frequencies on the human body (in Russian). Collection of scientific papers of the Novosibirsk Medical Institute, Novosibirsk, 1969.

Ostrovsky B. Flight 007 Mystery (in Russian). Publishing House "Nauka", Baku, 1997.

Ostrovsky B. The truth about the Bermuda Triangle (in Russian). Publishing House AST, Moscow, 2001.

Ostrovsky B. Bermuda Triangle: From Myths to Unraveling (in Russian). Publishing House "Astrel", Moscow, 2005.

Ostrovsky B. The legend of the three corners (in Russian). Journal "Itogi", December 6, 2010.

Ostrovsky B. Prisoner of the Mountain (in Russian). Journal "Itogi", July 16, 2012.

Ostrovsky B. Cold war in the air (in Russian). Newspaper "Nasha Versiya № 29 dated July 29, 2013.

Ostrovsky B. Psychosis of the planet Earth (in Russian). Publishing House AST, Moscow, 2014.

Ostrovsky B. Aircraft in the "Ring of Fire" (in Russian). Newspaper "Novaya Gazeta", March 24, 2014.

Ostrovsky B. Lost orientation (in Russian). Newspaper "Argumenty i fakty", March 26, 2014.

Ostrovsky B. Psychosis of the Earth (in Russian). Newspaper "Argumenty i fakty", 2015.

Ostrovsky B. Tectonic hypothesis of some air crashes (in Russian). Collection of scientific papers "Actual problems of science and education", Ministry of Education and Science of the Russian Federation. Zhukovsky, Russia, 2015, p.359.

Ostrovsky B. The Mystery of White Water (in Russian). Newspaper "Argumenty i fakty", October 30, 2019.

Petrushevsky B.A. Earthquakes and the possibilities of their prediction (in Russian). Moscow, 1961.

Polya G. Mathematics and Plausible Reasoning (in Russian). Moscow, 1976.

Presman A. Electromagnetic fields and wildlife (in Russian). Moscow, 1968.

Raitt H. Exploring the Deep Pacific (in Russian). Moscow, 1961.

Roberts E. Our quaking Earth (in Russian). Moscow, 1966.

Rousseau P. Les Tremblements de terre (in Russian). Moscow, 1966.

Salikhov et al. Experimental revealing of the response of lithospheric processes in background atmosphere in the period of intensification of seismic processes (in Russian). Institute of Ionosphere, National Center for Space Research and Technology, Almaty, Kazakhstan Republic, 2016.

Seabrook W. Robert Wood (in Russian). Moscow, 1946.

Skryagin L. Secrets of sea disasters (in Russian). Moscow, 1986.

Skryagin L. 300 catastrophes that shook the world (in Russian). Moscow, 1996.

Stakhanov I. On the physical nature of ball lightning(in Russian). Moscow, 1979.

Stevens S. Experimental psychology (in Russian). Vol.2, Moscow, 1963.

Subbotin S. Processes in the upper mantle of the earth (in Russian). Kiev, Ukraine,1964.

Taziev G. When the Earth quakes (in Russian). Moscow, 1968.

Ulomov V. Attention: Earthquake (in Russian). Tashkent, 1971.

Walter G. The living Brain (in Russian). Moscow, 1969.

Hoefling H. All the miracles in one book (in Russian). Moscow, 1983.

Kholodov Yu. Magnetism in biology (in Russian). Moscow, 1970.

Khorbenko I. Sound, ultrasound, infrasound (in Russian). Moscow, 1978.

Hofelman K. All disasters of the Universe (in Russian). Moscow, 2000.

Shuleikin Essays on the physics of the sea (in Russian). Moscow, 1962.

Shulman S. Aliens over Russia (in Russian). Moscow, 1990.